Studies in Logic
Logic and Cognitive Systems
Volume 46

Questions, Inferences,
and Scenarios

Volume 35
Logic is not Mathematical
Hartley Slater

Volume 36
Understanding Vagueness. Logical, Philosophical and Linguistic Perspectives
Petr Cintula, Christian G. Fermüller, Lluís Godo and Petr Hájek, eds.

Volume 37
Handbook of Mathematical Fuzzy Logic. Volume 1
Petr Cintula, Petr Hájek and Carles Noguera, eds.

Volume 38
Handbook of Mathematical Fuzzy Logic. Volume 2
Petr Cintula, Petr Hájek and Carles Noguera, eds.

Volume 39
Non-contradiction
Lawrence H. Powers, with a Foreword by Hans V. Hansen

Volume 40
The Lambda Calculus. Its Syntax and Semantics
Henk P. Barendregt

Volume 41
Symbolic Logic from Leibniz to Husserl
Abel Lassalle Casanave, ed.

Volume 42
Meta-argumentation. An Approach to Logic and Argumentation Theory
Maurice A. Finocchiaro

Volume 43
Logic, Truth and Inquiry
Mark Weinstein

Volume 44
Meta-logical Investigations in Argumentation Networks
Dov M. Gabbay

Volume 45
Errors of Reasoning. Naturalizing the Logic of Inference
John Woods

Volume 46
Questions, Inferences, and Scenarios
Andrzej Wiśniewski

Studies in Logic Series Editor
Dov Gabbay   dov.gabbay@kcl.ac.uk

Questions, Inferences, and Scenarios

Andrzej Wiśniewski

© Individual author and College Publications 2013.
All rights reserved.

ISBN 978-1-84890-120-9

College Publications
Scientific Director: Dov Gabbay
Managing Director: Jane Spurr

http://www.collegepublications.co.uk

Original cover design by Orchid Creative   www.orchidcreative.co.uk
Printed by Lightning Source, Milton Keynes, UK

All rights reserved. No part of this publication may be reproduced, stored in a retrieval system or transmitted in any form, or by any means, electronic, mechanical, photocopying, recording or otherwise without prior permission, in writing, from the publisher.

Contents

Part I Questions

Questions: an Informal Analysis 5
   1.1 Open-condition questions 5
   1.2 Delimited-condition questions 8
   1.3 Choice questions 9
   1.4 Topically-oriented questions................................. 10

Questions of Formal Languages 13
   2.1 Augmenting formal languages with questions.................. 13
   2.2 Questions vs. answers 14
   2.3 A semi-reductionistic approach 16
       2.3.1 "Knowing a question" 16
       2.3.2 A semi reduction to sets of answers 17
   2.4 Exemplary formal languages with questions 18
       2.4.1 $\mathcal{L}^?_{cpl}$: A propositional language with questions 18
       2.4.2 Modal propositional languages with questions 20
       2.4.3 $\mathcal{L}^?_{fom}$: A first-order language with questions 21
       2.4.4 $\mathcal{L}^?_{\vdash cpl}$: An erotetic sequent language 22

Elements of Minimal Erotetic Semantics: Declaratives 25
   3.1 Admissible partitions and entailment......................... 25
       3.1.1 Entailment 26
       3.1.2 The minimalistic method 26
       3.1.3 The direct method: language $\mathcal{L}^?_{\vdash cpl}$ 27
       3.1.4 The indirect method: language $\mathcal{L}^?_{cpl}$ 29
       3.1.5 The indirect method: language $\mathcal{L}^?_{S4}$ 30
       3.1.6 The indirect method: language $\mathcal{L}^?_{fom}$ 31
   3.2 Multiple-conclusion entailment 33
       3.2.1 Multiple-conclusion entailment vs. single-conclusion
               entailment 34
   3.3 Eliminating and narrowing down 34

Elements of Minimal Erotetic Semantics: Questions 37
    4.1   Soundness of a question 37
    4.2   Safety and riskiness 38
    4.3   Presuppositions and prospective presuppositions............... 39
    4.4   Normal questions and regular questions 41
    4.5   Self-rhetorical questions and proper questions 41
    4.6   Relative soundness 42
    4.7   Types of answers 43
        4.7.1   Just-complete answers 43
        4.7.2   Partial answers....................................... 43
        4.7.3   Eliminative answers 44
        4.7.4   Corrective answers 45
    4.8   The applicability issue 45

Part II Inferences

Inferential Erotetic Logic: An Introduction 49
    5.1   Erotetic inferences 49
    5.2   Validity of erotetic inferences 50
        5.2.1   Validity of erotetic inferences of the first kind 50
        5.2.2   Validity of erotetic inferences of the second kind........ 51
    5.3   Validity and question raising.................................. 56
    5.4   The logical basis of IEL 56
        5.4.1   Syntax 56
        5.4.2   Semantics 57

Evocation of Questions 59
    6.1   Definition of evocation 59
        6.1.1   Transmission of truth into soundness................... 60
        6.1.2   Informativeness 61
    6.2   Some properties of evocation.................................. 61
        6.2.1   A digression: Meheus' analysis 62
        6.2.2   Generation of questions 63
    6.3   Examples of evocation 63
    6.4   Evocation and validity 65
        6.4.1   Some comments 66

Erotetic Implication 67
    7.1   Definition of erotetic implication 67
        7.1.1   Eliminating a direct answer vs. narrowing down the set of direct answers 68
        7.1.2   Narrowing down vs. answering 70
        7.1.3   Some comparisons 70
    7.2   Erotetic implication and validity 72
    7.3   Some properties of erotetic implication 74
        7.3.1   Mutual soundness 74
        7.3.2   Monotony and transitivity issues 75

7.4		Some special kinds of erotetic implication 76
	7.4.1	Regular erotetic implication 76
	7.4.2	Strong erotetic implication 76
	7.4.3	Pure erotetic implication. Analyticity 76
7.5		Examples of erotetic implication 77
	7.5.1	Pure erotetic implication: examples 78
	7.5.2	Erotetic implication on the basis of non-empty sets of d-wffs: examples 80
7.6		Erotetic implication, evocation, and goal-directness 84
	7.6.1	Evocation as erotetic implication by non-factual questions 84
	7.6.2	Answering evoked questions by means of answers to implied questions 85

Socratic Transformations 89

8.1		Language $\mathcal{L}^?_{\vdash cpl}$ again 89
	8.1.1	Syntax 89
	8.1.2	Semantics 90
8.2		From questions to questions 91
	8.2.1	Some examples 91
	8.2.2	$\mathbb{E}^*$: An erotetic calculus for CPL 94
	8.2.3	Other erotetic calculi 97
	8.2.4	Erotetic calculi vs. sequent calculi 98
	8.2.5	Internal question processing 98

Part III Scenarios

E-scenarios 103

9.1		Erotetic Decomposition Principle 103
	9.1.1	Interrogative Model of Inquiry 103
9.2		E-scenarios: intuitions 105
	9.2.1	First story 105
	9.2.2	Second story 107
	9.2.3	Third story 109
9.3		E-scenarios: definitions 110
	9.3.1	Erotetic derivations 110
	9.3.2	E-scenarios as families of e-derivations 113
	9.3.3	E-scenarios as labelled trees 115
9.4		The Golden Path Theorem 116
9.5		A pragmatic account of e-scenarios 117
	9.5.1	Compression and conciseness 122
	9.5.2	Imperative counterparts of e-scenarios 124

Some Special Kinds of E-scenarios 127
10.1 Complete and incomplete e-scenarios 127
10.2 Pure e-scenarios and standard e-scenarios 127
  10.2.1 Standard e-scenarios for logical constants. The case of Classical Logic 128

        10.2.2 Standard decomposition e-scenarios 131
    10.3 Information-picking e-scenarios 133

Operations on E-scenarios 137
    11.1 Embedding 137
        11.1.1 A formal account of embedding 141
        11.1.2 A procedural account of embedding 145
    11.2 Contraction 146
        11.2.1 A formal account of contraction 147
        11.2.2 Examples of contraction 149

Querying Atomically 151
    12.1 Atomic e-scenarios for quantifier-free whether-questions 153
    12.2 Transforming an e-scenario into an atomic one 157

E-scenarios and Problem Solving 163
    13.1 Two kinds of problem decomposition 163
    13.2 Dynamic decomposition via e-scenarios 164
        13.2.1 Preliminary e-scenarios 164
        13.2.2 From query resolution to contraction 165
        13.2.3 Embedding as a rescue option 166
        13.2.4 Other rescue options and gains from a failure 168
        13.2.5 Fine-tuning and systematic embedding 169

Index 171

References 175

Preface

The importance of questions is beyond doubt. But the degree of attention paid to them in logic and linguistics is still less than they deserve. For decades research on questions focused on their representation as well as the answerhood problem. The priorities started to change in the 1980's. Generally speaking, research on how questions *function* (in inquiry, dialogues, reasoning, issue management, and so forth) gradually overshadowed research on what questions *are*. This change in perspective has been initiated by Jaakko Hintikka with his Interrogative Model of Inquiry.

The interest in questions and questioning is currently growing. In particular, questions became a full-fledged category in dynamic epistemic logic (cf. e.g., Minică (2011), van Benthem and Minică (2012), Peliš and Majer (2011)), and in belief revision theory (cf. Olsson and Westlund (2005), Enqvist (2010)). Theories of questions became indispensable constituents of dialogue theories (cf. Ginzburg (2012), Asher and Lascarides (2003)). Logic of questions attracted attention of the adaptive logic community (see e.g. Meheus (2001), De Clercq (2005), Batens (2007)). And, last but not least, research on questions is an important part of the inquisitive semantics programme (cf. e.g. Groenendijk and Roelofsen (2009), Groenendijk (2011)).

This book presents an *inferential* approach to the logic of questions.[1] The core part of it is *Inferential Erotetic Logic*, that is, to speak generally, a logic which analyses inferences which have questions as conclusions and gives an account of *validity* of these inferences. The idea originates from the late 1980's. The monograph Wiśniewski (1995) summarizes results obtained until the early nineties. Many things have happened since then. Although the core insights have remained unchanged, a more general account of Inferential Erotetic Logic was elaborated and some applications became known.

The book consists of three parts.

The chapters included in the second part, with the exception of the last one, provide an introduction to Inferential Erotetic Logic. An attempt was made to express the basic ideas as simply as possible. Moreover, the account presented is

---

[1] The logic of questions is sometimes labelled *erotetic logic*, from Greek "erotema" meaning "question".

more general than that taken in the 1995 monograph. In particular, we operate within a setting that does not presuppose Classical Logic and model-theoretic semantics.

The setting is described in the first part of the book. It is called *Minimal Erotetic Semantics*. The name is a telling one. The assumptions are really minimal, but still enable us to introduce some important concepts pertaining to questions and questioning.

As might be expected, the conceptual apparatus of Inferential Erotetic Logic has found successful applications in the area of problem solving. But, somewhat unexpectedly, the area of applicability has extended to proof theory. The last chapter of Part Two, and the whole of Part Three are devoted to these issues. Chapters 11 and 12 are rather technical due to the fact that they include some material that is published here for the first time. The remaining chapters are written in a more relaxed way. However, we always give the reader an indication as to where to find proofs and other technical details.

I am very grateful to Dorota Leszczyńska-Jasion, Paweł Łupkowski and Mariusz Urbański for their valuable comments which enabled me to improve earlier versions of this book. I am especially indebted to Jonathan Ginzburg for encouragement, criticism and help. Needless to say, all errors are mine.

Work on substantial parts of this book was possible due to the financial support of the National Science Council, Poland (DEC-2012/04/A/HS1/00715).

Poznań, June 2013

Part I

Questions

1
Questions: an Informal Analysis

Natural languages include questions of different kinds or types. The terminology pertaining to questions vary from theory to theory, however. According to Harrah (2002), the following labels for question-types are used by most theorists (examples are also taken from Harrah (2002), pp. 1–2):

LABEL	EXAMPLE
whether	Is two even or odd?
yes-no	Is two a prime number?
which	Which even numbers are prime?
what	What is Church's Thesis?
who	Who is Bourbaki?
why	Why does two divide zero?
deliberative	What shall I do now?
disjunctive	How long is your new proof, or do you have a shorter one?
hypothetical	If you had a proof, how long would it be?
conditional	If you now have a proof, how long is it?
given-that	Given that Turing's Conjecture is provable, is Church's Thesis provable?

The list is by no means exhaustive. One can easily add to it *when*, *where*, *how*, etc. Linguists tend to speak about *constituent*, *alternative*, and *polar* questions.

In this chapter we give an informal analysis of some basic categories of natural language questions. For convenience, we label the categories with descriptive names. These names are not commonly used in the literature. However, we chose them for a reason: it is better to use less theory-laden terms in a general presentation of an area.

1.1 Open-condition questions

There are questions which, generally speaking, express open conditions requested to be filled. An appropriately filled condition *contributes to* – and in some cases even *is* – an answer to the question which an idealized answerer would eventually give. Certain *who*, *where* and *when* questions, as well as some, but not all, *what*, *how* and *which* questions, are of this kind.

As an illustration, let us consider:

$$\text{Who went for a walk?} \tag{1.1}$$

$$\text{Where did Bill go?} \tag{1.2}$$

The open conditions are:

$$\text{... went for a walk} \tag{1.3}$$

$$\text{Bill went to ...} \tag{1.4}$$

But what does it mean that an open condition is appropriately filled?

Clearly, both "who" and "where" set the kinds of objects that are supposed to satisfy the relevant open conditions. Similarly for "when", "how often", "how many", "how much", "how far", "how long", "what time", "which boy", "which even number(s)", and so forth. So open conditions should be completed with expressions that refer to objects of the relevant categories: persons, places, etc. This, however, raises two issues.

First, *how many* persons/places – and in general: objects of the required kind – should be referred to in an answer? Depending on the solution, different readings of interrogative sentences emerge. The principal options are:[1]

OPTION	LABEL
only one	ONE-CASE reading
possibly multiple examples	CASE-OR-CASES reading
a complete list	ALL-THE-CASES reading

Consider the following statements:

$$\text{Bill went for a walk.} \tag{1.5}$$

$$\text{Bill, Mary and Harry went for a walk.} \tag{1.6}$$

$$\text{Bill, Mary, Harry, and only they went for a walk.} \tag{1.7}$$

When (1.1) is construed as one-case question, statement (1.5) is a possible answer to the question. Statements (1.5) and (1.6) can be regarded as examples of answers to (1.1) understood as case-or-cases question. Statement (1.7) is a possible answer to (1.1) read as all-the-cases question.

The second issue is: *how* persons/places – and in general: objects belonging to the relevant categories – should be referred to?

In particular, is giving a proper name necessary? Certainly not. Consider:

$$\text{Philosophers went for a walk.} \tag{1.8}$$

$$\text{Some drunk guys went for a walk.} \tag{1.9}$$

---

[1] These are qualitative options. Some theorists supplement them with certain quantitative options.

$$\text{Some drunk guys, and only them went for a walk.} \qquad (1.10)$$

What is more, one may wonder if responding to a who-question or a where-question with a name (of an object belonging to the relevant category) is always sufficient. For example, reply (1.5) to question (1.1) is a satisfactory answer only if it is known who Bill is. Similarly, the following reply to question (1.2):

$$\text{Bill went to Monclova.} \qquad (1.11)$$

is a satisfactory answer only if it is known what Monclova is.

The situation is analogous in the case of other questions of the analysed kind.

Some semantic constraints. It cannot be said that every question is subjected to each of the understandings sketched above. Let us consider:

$$\text{Who is Bill's biological father?} \qquad (1.12)$$

A man has exactly one biological father and thus only the one-case reading is possible.

As for (1.12), some meaning component of the condition fixes the reading. Sometimes, however, the effect is due to the interrogative phrase only. Here is an example:

$$\text{How many students has he got?} \qquad (1.13)$$

Finally, the grammar can exclude the one-case reading. Please consider:

$$\text{Which even numbers are prime?} \qquad (1.14)$$

Context. A natural-language question is always asked in a context: social, theoretical, of an explanation previously given, of a story just told, etc. Thus although questions permit diverse readings, the context usually disambiguates them. The disambiguation pertains both to the "how many" issue and the "how-to-refer-to" issue. For instance, question (1.1) asked by a policeman investigating a criminal case is most likely to be understood as an all-the-cases question where the relevant persons should be referred to with their first and last names.

If needed, the disambiguation effect can also be achieved by linguistic means. The simplest way is to add an imperative sentence to an interrogative one. For instance:

$$\text{Who went for a walk? Please give the first and last name}$$
$$\text{of each of them.} \qquad (1.15)$$

Multiple wh-questions. The picture gets more complicated when multiple wh-questions are taken into consideration. Let us consider:

$$\text{Who likes whom?} \qquad (1.16)$$

$$\text{Which boys love which girls?} \qquad (1.17)$$

For simplicity, assume that it is sufficient to refer to boys and girls with their first names. Still, it is unclear whether an answer to (1.17) is supposed to specify a list of boys together with associated sub-lists of girls (namely, girls loved by a given boy), or is supposed to specify a list of "boy-girl" couples. Moreover, the "how much" issue remains open, also at the conceptual level.[2]

1.2 Delimited-condition questions

There are questions that express a condition to be filled, yet associated with a list of instances. These instances delimit the condition by specifying the relevant options.

Consider the following:

Who went for a walk: Bill, Mary, or Harry? (1.18)

Where did Bill go: somewhere in Poland, or to Siberia? (1.19)

The condition included in (1.18) is supposed to be filled by "Bill", "Mary", or "Harry"; the results constitute the "alternatives" offered by the question. Thus the "how-to-refer to" issue does not arise. However, (1.18) can still be construed as one-case, or case-or-cases, or all-the-cases question. Question (1.19) can be viewed analogously.

Now let us consider:

Which is Bill's favourite painter: Matisse or Cezanne? (1.20)

Since both "favourite" and "which" presuppose uniqueness, only the one-case reading is possible. Similarly, the following:

What did Bill do: stayed at home or went for a walk? (1.21)

permits only the one-case reading. Again, the reason is semantic, though a different one than before: one cannot stay at home and to go for a walk at the same time. In other words, the "alternatives" offered by the question exclude each other.

Delimitation by context. It often happens that the meaning of an interrogative sentence that primarily expresses an open-condition question is the same as the meaning of the corresponding delimited-condition question. For instance, if you were just told that Andrew left for Paris, London, or Moscow, you would construe the interrogative sentence "Where did Andrew leave for?" as a delimited-condition question. Much more can be said about how contextual factors determine meanings of some interrogative sentences[3], but since this chapter is of an introductory character, we will not address this issue more extensively here.

---

[2] For multiple wh-questions see, e.g., Kubiński (1971), Kubiński (1980), Higginbotham and May (1981), Ginzburg and Sag (2000).

[3] Cf. e.g. Ginzburg (1995).

1.3 Choice questions

Some questions do not involve any explicit conditions in their contents, but, instead, list certain "alternatives" among which a choice is requested to be made. Let us consider the following yes-no question:

$$\text{Did Bill go for a walk?} \tag{1.22}$$

The most predictable replies are "Yes" and "No". The first corresponds to the alternative explicitly listed:[4]

$$\text{Bill went for a walk.} \tag{1.23}$$

However, the alternative that corresponds to "No" is not explicitly listed. As for the example considered, "No", depending on the context (and possibly some pragmatic factors, such as the accentuation in which (1.22) is uttered) means either:

$$\text{It is not the case that Bill went for a walk.} \tag{1.24}$$

or:

$$\text{It was not Bill, but someone else who went for a walk.} \tag{1.25}$$

or:

$$\text{Bill did something else than to go for a walk.} \tag{1.26}$$

or even:

$$\text{Bill went, but not for a walk.} \tag{1.27}$$

Thus the category of yes-no questions is not homogeneous.[5]

A whether-question explicitly lists all the relevant alternatives. When only two alternatives are listed, these alternatives usually exclude each other, as in:

$$\text{Is Bill a genius, or an idiot?} \tag{1.28}$$

but this is not a rule, viz.:

$$\text{Is Bill a logician, or a philosopher?} \tag{1.29}$$

Question (1.29) permits the one-case, case-or-cases, and all-the-cases readings. The latter licenses the following choices: "Bill is a logician but not a philosopher", "Bill is a philosopher but not a logician" and "Bill is both a philosopher and a logician". The case-or-cases reading licenses: "Bill is a logician", "Bill is a philosopher" and "Bill is both a logician and a philosopher".

---

[4] In English a simple grammatical transformation is needed. In Polish even this is not necessary. The Polish translation of (1.22) is "Czy Bill poszedł na spacer?", and of (1.23) is "Bill poszedł na spacer."

[5] An analogy to the *de dicto – de re* distinction known from modal logic suggests itself. When the second alternative is a sentential negation of the alternative explicitly listed, the question is *de dicto*; when only a (proper) constituent of the first alternative is "negated", the question is *de re*. Of course, the subcategory of *de re* yes-no questions splits further, as the above example illustrates.

Choice questions often have their direct counterparts among delimited-condition questions, and vice-versa. Yet this is not a general rule. Here are counter-examples.

$$\text{Has Bill left for a while, or has he never lived here?} \quad (1.30)$$

$$\text{Is Bill a philosopher, or a poet, or does he only like to} \\ \text{impress people with clumsy statements?} \quad (1.31)$$

Finally, let us consider the following interrogative sentence:

$$\text{Is Bill happy and rich?} \quad (1.32)$$

(1.32) can be construed either as a (*de dicto*) yes-no question or as a *conjunctive question*. In the latter case the following choices are licensed: "Bill is happy and rich", "Bill is happy but not rich", "Bill is not happy, but he is rich", and "Bill is neither happy nor rich".

1.4 Topically-oriented questions

The last category of natural language question we are going to distinguish here is, at first sight, very heterogeneous. Any of the following is a question belonging to the category:

$$\text{Why did Bill divorce for the first time?} \quad (1.33)$$
$$\text{Why are logicians more handsome than philosophers?} \quad (1.34)$$
$$\text{Why does copper turn green when exposed to air?} \quad (1.35)$$
$$\text{How are you going to resolve this problem?} \quad (1.36)$$
$$\text{How did it happen that Bill became a logician?} \quad (1.37)$$
$$\text{What do you know about the case?} \quad (1.38)$$

What do the above questions have in common? First, some negative property: none of them specifies, directly or even indirectly, the list of relevant alternatives. Second, they display some features (see below) which differentiate them from open-condition questions analysed in section 1.1. Third, all of them can be informally described as *topically-oriented*.

As for why-questions, one could have analysed them as expressing a condition of the form:

$$\varrho \;\; \text{because} \ldots \quad (1.39)$$

(where ϱ stands for the declarative sentence that "occurs" in a question[6]) requested to be filled. Yet, the mere "why" does not set the kind of objects that are supposed to satisfy condition (1.39). The meaning of "because" depends on the meaning of ϱ and, what is more important, in order to have an account of

---

[6] We have used quotation marks, because in English a simple grammatical transformation is required. In Polish no transformation is necessary; why-questions have the form "Dlaczego ϱ?".

possible meanings one needs tools stronger than these offered by logic or linguistics themselves. The so-called models of explanation are illuminating here, but they lie outside logic or linguistics proper. The models determine how to appropriately fill condition (1.39).[7] It is clear, however, that explanation of actions, of facts, and of laws diverge. Moreover, contemporary philosophy of science offers many models of each of these kinds of explanation. The only general claim concerning why-questions is: they are topically-oriented, where the topic is an explanation.

The situation with the remaining questions cited above is even worse. One would not achieve much progress by analysing them as requests to fill the open conditions:

I am going to resolve this problem in the following way: ... (1.40)

... and then Bill became a logician. (1.41)

This is what I know about the case: ... *and* ... *and* ... (1.42)

What can be said by now is: they are topically-oriented, where the topics are a way of resolving a problem, the genesis of an event, and knowledge about a case, respectively.

Needless to say, topically-oriented questions constitute a hard nut for the analysis of questions, both logical and linguistic.

Remarks. It cannot be said that the categories distinguished above cover the whole realm of questions. We did not aim at completeness, however. Neither we aimed at exactness. Even the degree of originality of the proposed account is, say, moderate: we relied upon ideas present in various theories. But our goals were different. Besides obvious introductory purposes, we wanted to illustrate two claims. First, the area of questions is far from being homogeneous. Second, similarly as in the cases of declaratives and imperatives, meanings of interrogative sentences are often co-determined by contextual/pragmatic factors. As a matter of fact, these are the main reasons for which providing an adequate formal analysis of questions still constitutes a difficult task.

A brief historical digression. In the late fifties/early sixties of the 20th century the conceptual apparatus of modern formal logic begun to be extensively applied in the area of questions and questioning. Gerold Stahl, Tadeusz Kubiński, David Harrah, Nuel D. Belnap, and Lennart Åqvist established the first widely elaborated logical theories of questions.[8] These theories disagreed in many respects. A lot of things happened since then, new theories were proposed and linguists entered the game. What has remained unchanged, however, is the lack of agreement concerning the basic concepts and, what is more striking, the general perspective.

The survey paper Harrah (2002) provides a comprehensive exposition of logical theories of questions elaborated till late 1990s. Supplementary information

---

[7] For this approach to why-questions see e.g. Kuipers and Wiśniewski (1994), Wiśniewski (1999), Grobler and Wiśniewski (2005), Grobler (2006).
[8] The story is told in detail in Harrah (1997).

about more linguistically oriented approaches can be found, e.g., in Groenendijk and Stokhof (1997) (reprinted as Groenendijk and Stokhof (2011)) and Krifka (2011). The paper Ginzburg (2011) provides a survey of recent developments in the research on questions, both in logic and in linguistics. A general overview of approaches to questions and their semantics can also be found in Wiśniewski (201x*b*).

2

Questions of Formal Languages

2.1 Augmenting formal languages with questions

A logician interested in questions and questioning most often starts with a formal (or formalized) language which initially does not contain direct counterparts of questions. Generally speaking, questions/interrogatives can be incorporated into a formal language in two ways.

1. (The "DEFINE WITHIN" approach.) One can *embed* questions into a language. To be more precise, one can regard as questions some already given well-formed formulas differentiated by some semantic feature(s), or construe questions as meanings of some specific (but already given) well-formed formulas.

This way of proceeding is natural when the so-called paraphrase approach to questions is adopted, that is, it is believed that the meaning of an interrogative sentence can be adequately characterized by a paraphrase that specifies the typical use of the sentence or the relevant illocutionary act performed in uttering the sentence. For example, one can claim that the following:

$$\textit{Does Bill like Mary?} \tag{2.1}$$

is synonymous with:

$$\textit{Bring it about that I know whether Bill likes Mary.} \tag{2.2}$$

or with:

$$\textit{I request that you assert that Bill likes Mary} \\ \textit{or deny that Bill likes Mary.} \tag{2.3}$$

The paraphrase (2.2) can be formalized within a setting which involves epistemic operators and imperative operator(s). (2.3), in turn, can be formalized within a logical theory of illocutionary acts.[1] In both cases no separate, primitive syntactic category of interrogatives is needed. For convenience, one can then define "interrogative formulas", but they will be only abbreviations of their counterparts, and, what is more important, their semantics is just the semantics of the relevant well-formed formulas.

---

[1] (2.2) is a Hintikka-style paraphrase (cf. e.g. Hintikka (1976), Hintikka (1978)), while (2.3) agrees with Vanderveken's-style approach to questions (cf. Vanderveken (1990)).

2. (The "ENRICH WITH" approach.) One can *enrich* a language with questions/interrogatives. In order to achieve this, one adds to the vocabulary some question-forming expressions and then introduces questions/interrogatives syntactically, as a new category of well-formed formulas. The new category is disjoint with the remaining categories. This way of proceeding is natural when questions are conceived in accordance with the independent meaning thesis, according to which the meaning/semantic content of an interrogative sentence cannot be adequately characterized in terms of semantics of expressions that belong to other categories.

The second approach is sometimes called the *method of interrogative extensions*.[2]

Regardless of which approach is adopted, one ends with a class of *erotetic formulas* or *e-formulas* for short.

E-formulas of a formal language can be identified with *questions* of the language. This does not presuppose that questions are *defined* in purely syntactic terms. They can be characterized in semantic terms, as the well-formed formulas that "correspond" to questions semantically construed or have the semantic property of being a question. In general, the sets of e-formulas and declarative well-formed formulas (*d-wffs* for short) need not be disjoint. Usually they overlap when the "define within" approach is adopted, and are disjoint otherwise.

Terminology. When referring to natural-language questions we will be using the acronym NLQ (after *natural-language question*).

2.2 Questions vs. answers

Characterizing e-formulas is only the first step. The crucial point is to give an account of answers to them. This can be (and is!) done in many ways and by different means. But, of course, not completely arbitrarily. E-formulas are supposed to represent (at least some) NLQ's and answers to e-formulas should formalize/represent *possible* answers to the corresponding NLQ's. Usually, one aims at formalizing/representing *potentially resolving answers* to NLQ's. Answers of this kind are called, depending on a theory, *direct*, or *conclusive*, or *proper*, or *sufficient*, or *exhaustive*, or *complete*, or *congruent*, etc. For brevity, let us use "principal possible answer" (ppa for short) as a cover term for answers of the above kinds.

The concept of potentially resolving answer to a NLQ happens to be construed differently in diverse theories. It is also vague because the underlying intuitions are expressed by using, among others, pragmatic terms, e.g.:

- Harrah: a *direct answer* "gives exactly what the question calls for. (...) The label 'direct' (...) connotes both logical sufficiency and immediacy" (Harrah (2002), p. 1);

---

[2] The method was introduced by Tadeusz Kubiński in the 1950s (see Kubiński (1960) and Harrah (1997)).

- Belnap: *direct answers* "are directly and precisely responsive to the question, giving neither more nor less information than what is called for" (Belnap (1969), p. 124);
- Kubiński: *direct answers* are "these sentences which everybody who understands the question ought to be able to recognize as the simplest, most natural, admissible answers to the question" (Kubiński (1980), p. 12);
- Hintikka: a *potential conclusive answer* is "an answer which would satisfy the questioner *if* it were true and *if* he were in a position to trust the answer. By a *conclusive answer*, I mean a reply which does not require further backing to satisfy the questioner." (Hintikka (1978), p. 287).

On the other hand, NLQ's permit multiple readings. Or, to put it differently, in many cases contextual and/or pragmatic factors co-determine what is "directly and precisely responsive to the question, giving neither more nor less information than what is called for", or what is a just-sufficient (i.e. immediate and sufficient) possible answer, etc.

In any case, logical theories of questions do assign ppa's to e-formulas, and regard e-formulas as formalizations of NLQ's. So what one really gets is:

(♠) *An e-formula Q represents a NLQ Q^** CONSTRUED IN SUCH A WAY *that possible answers to Q^* having the desired semantic and/or pragmatic properties are represented/formalized by ppa's to Q.*

Example 2.1. Let the desired property be *just-sufficiency*, i.e. sufficiency conjoined with immediacy.

- Suppose that, as the outcome of an analysis, we get an e-formula whose set of ppa's is:
$$\{\mathsf{Pb}, \mathsf{Tb}\} \qquad (2.4)$$
where P and T are syntactically distinct one-place predicates and b is an individual constant. The e-formula represents a "choice" NLQ whose set of possible just-sufficient answers consists of two syntactically distinct sentences made up of a predicate and a proper name, where the predicates are applied to the name.
- The set of ppa's equals:
$$\{\mathsf{Pb}, \mathsf{Tb}, \mathsf{Pb} \wedge \mathsf{Tb}\} \qquad (2.5)$$
The corresponding e-formula formalizes a NLQ which differs from that formalized by the previous e-formula in having Pb∧Tb as the additional possible just-sufficient answer.
- The set of ppa's is:
$$\{\mathsf{Pb} \wedge \neg \mathsf{Tb}, \mathsf{Tb} \wedge \neg \mathsf{Pb}\} \qquad (2.6)$$
The corresponding e-formula represents a NLQ which, generally speaking, requests for an exclusive choice.
- The set of ppa's equals:
$$\{\mathsf{Pb}, \mathsf{Tb}, \neg(\mathsf{Pb} \vee \mathsf{Tb})\} \qquad (2.7)$$

Now, besides Pb and Tb, "neither" (i.e. ¬(Pb ∨ Tb)) is also a possible just-sufficient option.

- Finally, the set of ppa's is:

$$\{\text{Pb} \land \text{Tb}, \text{Pb} \land \neg \text{Tb}, \neg \text{Pb} \land \text{Tb}, \neg \text{Pb} \land \neg \text{Tb}\} \quad (2.8)$$

The relevant e-formula corresponds to a "partition" NLQ.

2.3 A semi-reductionistic approach

2.3.1 "Knowing a question"

The assignment of ppa's to e-formulas can be performed in many ways. Yet, simplicity is always a virtue, and a purely syntactic approach allows us to gain it for a low price. For this reason we adopt here a syntactic account.

A possible objection is: ppa's are supposed to satisfy some conditions which are expressed, inter alia, in pragmatic terms (see above), and the satisfaction of the relevant conditions is not a matter of syntax. For example, assume that ppa's are supposed to be the possible just-sufficient answers. Clearly, there are cases in which it is strongly context-dependent what sentence may be counted as a possible and just-sufficient answer to a NLQ. Moreover, there are NLQ's for which it makes no sense at all to speak about predetermined *sets* of possible just-sufficient answers; some why-questions and how-questions are often recalled in this context.

However, the above objection is not irrefutable. The following should be carefully distinguished: (a) ppa's to NLQ's, and (b) ppa's to e-formulas. If the latter are defined purely syntactically, one can still claim that an e-formula represents, in the sense described by (♠) above, NLQ(s). If a NLQ has many readings, it has many representations. The richer the formal language is, the more we can represent in it.

In view of the account of representation sketched above the famous Hamblin postulate:[3]

H$_2$: *Knowing what counts as an answer is equivalent to knowing the question.*

splits into:

H$_{2_1}$: *Knowing a NLQ is equivalent to knowing the e-formula that represents it.*

H$_{2_2}$: *Knowing the e-formula is equivalent to knowing what counts as ppa's to it.*

Thus "knowing a question" often yields a disambiguation.[4]

---

[3] Cf. Hamblin (1958), p.162.

[4] The concept of representation and the above schema can also be used when ppa's to questions of formal languages are not defined purely syntactically.

2.3.2 A semi-reduction to sets of answers

Since ppa's are what really count, a possibility, and an appealing one, is to construct e-formulas according to the following schema:

$$?\Theta \qquad (2.9)$$

where Θ is an expression of the object-level formal language such that Θ is equiform with the expression of the metalanguage which, in turn, designates the set of ppa's to the e-formula. For example, when we add the question mark ? and the brackets: $\{,\}$ to the vocabulary, we can enrich the language with e-formulas of the form:

$$?\{A_1, \ldots, A_n\}$$

where $n > 1$ and $A_1, \ldots, A_n$ are pairwise syntactically distinct declarative well-formed formulas of the initial language; these formulas are the ppa's to the e-formula.

What we have achieved is a kind of *semi-reduction* of questions to sets of declaratives. Let us stress that e-formulas are still expressions of an object-level language. In particular, $?\{A, B\} \neq ?\{B, A\}$.

An advantage of this semi-reductionistic approach is that it is now extremely easy to say what counts as a ppa to an e-formula, and what NLQ's are represented by the e-formula. A disadvantage is the lack of conciseness in some cases. However, we can always introduce abbreviations as a remedy to it.

Example 2.2. Consider the interrogative sentence:

$$\textit{Who is Bill: a philosopher or a theologian?} \qquad (2.10)$$

Clearly, different readings of (2.10) are allowed. Let Pb stand for "Bill is a philosopher", and Tb for "Bill is a theologian". Suppose that the required property of ppa's to NLQ's is just-sufficiency. The following e-formulas correspond to some of the readings (cf. Example 2.1 for details):

$$?\{\text{Pb}, \text{Tb}\} \qquad (2.11)$$

$$?\{\text{Pb}, \text{Tb}, \text{Pb} \wedge \text{Tb}\} \qquad (2.12)$$

$$?\{\text{Pb} \wedge \neg \text{Tb}, \text{Tb} \wedge \neg \text{Pb}\} \qquad (2.13)$$

$$?\{\text{Pb}, \text{Tb}, \neg(\text{Pb} \vee \text{Tb})\} \qquad (2.14)$$

$$?\{\text{Pb} \wedge \text{Tb}, \text{Pb} \wedge \neg \text{Tb}, \neg \text{Pb} \wedge \text{Tb}, \neg \text{Pb} \wedge \neg \text{Tb}\} \qquad (2.15)$$

(2.15) can be abbreviated as:

$$?\pm|\text{Pb}, \text{Tb}| \qquad (2.16)$$

2.4 Exemplary formal languages with questions

In this book we adopt the "enrich with" approach to questions of formal languages. Below we present examples of languages in which e-formulas occur. These languages, with the exception of the last one, are built according to the semi-reductionistic pattern described above. The last language is constructed differently, but still has a rather simple syntax.

Terminology: questions and direct answers. From now on, principal possible answers (ppa's) will be called *direct answers*. Direct answers to e-formulas will be defined syntactically. As for NLQ's, direct answers/ppa'a are supposed to be the *possible just-sufficient answers*, where "just-sufficient" means "satisfies the request of a question by providing neither less nor more information than it is requested". An e-formula Q is supposed to represent, in the sense specified by (♠) (see page 15), the corresponding NLQ(s).

For purely stylistic reasons, e-formulas of formal languages will be called below *questions* of these languages.

We use Q, Q^*, Q_1, ... as metalinguistic variables for questions, and A, B, C, D, possibly with subscripts, as metalinguistic variables for declarative well-formed formulas (d-wffs). The context will always decide what language we have in mind.

Notation. dQ stands for the set of direct answers to question Q.

In the metatheory we assume a version of set theory which incorporates not only sets, but also classes. We use standard set-theoretical terminology and notation. The expression "iff" abbreviates "if and only if".

2.4.1 $\mathcal{L}^?_{cpl}$: A propositional language with questions

Let $\mathcal{L}_{cpl}$ be the language of Classical Propositional Logic (CPL for short) with $\neg$ (negation), $\rightarrow$ (implication), $\wedge$ (conjunction), and $\vee$ (disjunction) as primitive connectives; for simplicity, we do not consider equivalence as a primitive connective of $\mathcal{L}_{cpl}$. *Well-formed formulas* (wffs for short) of $\mathcal{L}_{cpl}$ are defined as usual. We use p, q, r, s, p_1, ... for propositional variables. We adopt the usual conventions for omitting parentheses in wffs of $\mathcal{L}_{cpl}$. The equivalence connective $\leftrightarrow$ is defined in the standard way, viz.:

$$(A \leftrightarrow B) =_{df} (A \rightarrow B) \wedge (B \rightarrow A)$$

Now we construct a second language, labelled $\mathcal{L}^?_{cpl}$. The vocabulary of $\mathcal{L}^?_{cpl}$ includes the vocabulary of $\mathcal{L}_{cpl}$, the following signs: ?, {, }, and the comma. The language $\mathcal{L}^?_{cpl}$ has two categories of well-formed expressions: declarative well-formed formulas (d-wffs for short) and erotetic formulas, that is, questions.

A *d-wff* of $\mathcal{L}^?_{cpl}$ is simply a wff of $\mathcal{L}_{cpl}$.

A *question* of $\mathcal{L}^?_{cpl}$ is an expression of the form:

$$?\{A_1, \ldots, A_n\} \tag{2.17}$$

where $n > 1$ and $A_1, \ldots, A_n$ are nonequiform, that is, pairwise syntactically distinct, d-wffs of $\mathcal{L}^?_{cpl}$ (i.e. CPL-wffs). If $?\{A_1, \ldots, A_n\}$ is a question, then each of the d-wffs $A_1, \ldots, A_n$ is called a *direct answer* to the question, and these are the only direct answers to the question. Note that any question of $\mathcal{L}^?_{cpl}$ has at least two direct answers. Observe that the set of direct answers to a question of $\mathcal{L}^?_{cpl}$ is always finite.

Any question of the form (2.17) can be read:

Is it the case that A_1, or $\ldots$, or is it the case that A_n?

However, sometimes a different reading is available. The schema (2.17) is general enough to capture most (if not all) of propositional questions studied in the literature.

For example, a *simple yes-no question*, that is, a question whose set of direct answers consists of a sentence and its classical negation, can be formalized as a question of the form:

$$?\{A, \neg A\} \qquad (2.18)$$

and read:

Is it the case that A?

The d-wffs A and $\neg A$ are the *affirmative answer* and the *negative answer* to (2.18), respectively.

Conditional yes-no questions with irrevocable antecedents have the form of:

$$?\{A \wedge B, A \wedge \neg B\} \qquad (2.19)$$

whereas *conditional yes-no questions with revocable antecedents* are formalized by:

$$?\{A \wedge B, A \wedge \neg B, \, A\} \qquad (2.20)$$

A question of the form (2.19) can be read:

It is the case that A; is it also the case that B?

As far as questions of the form (2.20) are concerned, the recommended reading is:

Is it the case that A?; if so, is it also the case that B?

Questions falling under the schema:

$$?\{A \wedge B, A \wedge \neg B, \neg A \wedge B, \neg A \wedge \neg B\} \qquad (2.21)$$

can be read:

Is it the case that A and is it the case that B?

They may be called (binary) *conjunctive questions*.

For the sake of concision we adopt some notational conventions pertaining to questions which we will frequently refer to. Questions of the form (2.21) will be abbreviated as:

$$?\pm|A, B| \qquad (2.22)$$

Simple yes-no questions (i.e. questions of the form (2.18)) will be concisely written as:

$$?A \qquad (2.23)$$

Further notational conventions will be introduced when needed.

2.4.2 Modal propositional languages with questions

Modal propositional languages with questions are constructed similarly as the language $\mathcal{L}^?_{cpl}$ described above. The difference lies in taking the language $\mathcal{L}_\mu$ of a modal propositional logic μ as the point of departure.

To be more precise, we enrich the vocabulary of $\mathcal{L}_{cpl}$ with the modal operators $\Diamond$ of possibility and/or $\square$ of necessity.[5] As a result, we get the vocabulary of the language, $\mathcal{L}_\mu$, of a modal propositional logic μ. Wffs of $\mathcal{L}_\mu$ are defined in the standard manner. Then we build the language $\mathcal{L}^?_\mu$. We enrich the vocabulary of $\mathcal{L}_\mu$ with the signs: ?, {, }, and the comma. The d-wffs of $\mathcal{L}^?_\mu$ are the wffs of $\mathcal{L}_\mu$. Questions of $\mathcal{L}^?_\mu$ are expressions of the language falling under the schema (2.17) specified above; direct answers are characterized analogously. The conventions introduced in the previous section apply accordingly.

Let us consider questions of the form:

$$?\{A, \Diamond A, \square A, \neg A, \Diamond \neg A, \square \neg A\} \qquad (2.24)$$

where A is a CPL-wff, and $\Diamond$ as well as $\square$ are understood as alethic modalities. Some theorist claim that ppa's to polar questions comprise not only a sentence and its negation, but also the relevant "modalized" statements.[6] Questions of the form (2.24) correspond to polar questions construed that way.[7]

Note that the transition from (2.18) to (2.24) fits a certain general pattern: when we have a question falling under the schema:

$$?\{A_1, \ldots, A_n\} \qquad (2.17)$$

we also have the corresponding question of the form:

$$?\{A_1, \Diamond A_1, \square A_1, \ldots, A_n, \Diamond A_n, \square A_n\} \qquad (2.25)$$

However, $\Diamond$ and $\square$ need not be understood as alethic modalities. It is, in a sense, natural to construe "It is possible that" occurring in an answer as an epistemic modality, for example as "It is not ruled out" or "It is thought as possible". Under such a reading a modalized counterpart of a question of the form (2.17) of $\mathcal{L}^?_{cpl}$ falls under the schema:

$$?\{A_1, \Diamond A_1, \ldots, A_n, \Diamond A_n\} \qquad (2.26)$$

rather than under the schema (2.25).

---

[5] For simplicity, we do not consider the multimodal case here.
[6] Cf. Ginzburg (1995).
[7] Observe that when the underlying modal logic is only K, then A, $\Diamond A$ and $\square A$ are independent from each other, and similarly for their counterparts involving negation.

2.4.3 $\mathcal{L}^?_{fom}$: A first-order language with questions

Now let us take the language of Monadic First-Order Logic with Identity as the starting point. We designate this language by $\mathcal{L}_{fom}$. For simplicity, we assume that the vocabulary of $\mathcal{L}_{fom}$ contains an infinite list of individual constants, but does not contain function symbols. *Well-formed formulas* (wffs) of $\mathcal{L}_{fom}$ are defined in the standard way. By *terms* of $\mathcal{L}_{fom}$ we mean individual variables and individual constants of the language. Freedom and bondage of variables are defined as usual. A *sentential function* is a wff in which free variables occur; otherwise a wff is a *sentence*.

We construct a second language, $\mathcal{L}^?_{fom}$, which has a declarative part and an erotetic part. The vocabulary of $\mathcal{L}^?_{fom}$ consists of the vocabulary of $\mathcal{L}_{fom}$ and the following signs: ?, {, }, **S**, **U**, and the comma.

The declarative part of $\mathcal{L}^?_{fom}$ is the language $\mathcal{L}_{fom}$ itself. *Declarative well-formed formulas* (d-wffs) of $\mathcal{L}^?_{fom}$ are the wffs of $\mathcal{L}_{fom}$, and similarly for other concepts introduced above.

As far as the erotetic part of $\mathcal{L}^?_{fom}$ is concerned, we have three categories of questions.

Whether-questions of $\mathcal{L}^?_{fom}$ fall into the schema (2.17) specified in section 2.4.1, that is, are of the form:

$$?\{A_1, \ldots, A_n\}$$

where $n > 1$ and $A_1, \ldots, A_n$ are nonequiform (i.e. pairwise syntactically distinct) sentences of $\mathcal{L}^?_{fom}$. Direct answers to questions of the form (2.17) of $\mathcal{L}^?_{fom}$ are defined as above. We adopt analogous notational conventions as in the case of $\mathcal{L}^?_{cpl}$.

Moreover, $\mathcal{L}^?_{fom}$ contains questions falling under the schemata:

$$?\mathbf{S}(Ax) \qquad (2.27)$$

$$?\mathbf{U}(Ax) \qquad (2.28)$$

where x stands for an individual variable and Ax is a sentential function of $\mathcal{L}^?_{fom}$ which has x as the only free variable.

Questions of the form (2.27) can be read:

Which x is such that Ax?

We shall call them *existential which-questions*.

By a direct answer to an existential which-question $?\mathbf{S}(Ax)$ we mean a sentence of $\mathcal{L}^?_{fom}$ which is an instantiation (by an individual constant) of the sentential function Ax. Thus direct answers to a question of the form (2.27) are sentences of the form $A(x/c)$, where c is an individual constant.

Questions having the form of (2.28) can be read:

What are all of the x's such that Ax?

We call them *general which-questions*.

By a direct answer to a general which-question $?\mathbf{U}(Ax)$ we mean a sentence of $\mathcal{L}^?_{fom}$ falling under the schema:

$$A(x/c_1) \land \ldots \land A(x/c_n) \land \forall x(Ax \to x = c_1 \lor \ldots \lor x = c_n) \qquad (2.29)$$

where $n \geq 1$ and $c_1, \ldots, c_n$ stand for distinct individual constants.

The symbols $\mathbf{S}$ and $\mathbf{U}$ belong to the vocabulary of the object-level language $\mathcal{L}^?_{fom}$. However, we can introduce them to the metalanguage as well (but with different meanings). We can assume that on the metalanguage level $\mathbf{S}(Ax)$ designates the set of all the sentences of the form $A(x/c)$, whereas $\mathbf{U}(Ax)$ designates the set of all the sentences of the form (2.29). Now we are justified in saying that each question of $\mathcal{L}^?_{fom}$ consists of the sign ? followed by an (object-level language) expression which is equiform to a metalanguage expression that designates the set of direct answers to the question. In other words, the semi-reductionistic approach is retained: questions of $\mathcal{L}^?_{fom}$ fall under the schema (2.9) (see page 17).

The expressive power of $\mathcal{L}^?_{fom}$ is limited. In particular, only some open-condition questions (see Chapter 1, section 1.1) are represented by questions of $\mathcal{L}^?_{fom}$. However, we aim here at an illustration rather than at generality. And nothing prevents us from taking a richer first-order language (or a higher-order language) as the point of departure, and from introducing other categories of wh-questions according to the semi-reductionistic pattern. For possible developments see Wiśniewski (1995), Chapter 3.

2.4.4 $\mathcal{L}^?_{\vdash cpl}$: An erotetic sequent language

Now let us present an example of a language especially designed to tackle with a certain class of problems, namely logical problems of entailment/derivability, validity, and inconsistency. For simplicity, we focus on the (classical) propositional case; when more sophisticated cases are analysed, the construction goes along similar lines, but is more complex.[8]

As in section 2.4.1, we take the language $\mathcal{L}_{cpl}$ of CPL as the point of departure, and then we construct a language, $\mathcal{L}^?_{\vdash cpl}$, which has a declarative part and an erotetic part. However, both d-wffs of $\mathcal{L}^?_{\vdash cpl}$ and questions of $\mathcal{L}^?_{\vdash cpl}$ are defined differently than in the case of $\mathcal{L}^?_{cpl}$.

The vocabulary of the language $\mathcal{L}^?_{\vdash cpl}$ contains the vocabulary of $\mathcal{L}_{cpl}$ and the following signs: ?, $\vdash$, ng ($\mathcal{L}^?_{\vdash cpl}$-negation), & ($\mathcal{L}^?_{\vdash cpl}$-conjunction), as well as the comma.

Now we introduce the concept of a $\mathcal{L}_{cpl}$-sequent. By a $\mathcal{L}_{cpl}$-*sequent* we mean an expression of the form:

$$S \vdash A \qquad (2.30)$$

where A is a (single!) wff of $\mathcal{L}_{cpl}$ (that is, a CPL-formula), and S is a finite, possibly empty, sequence of wffs of $\mathcal{L}_{cpl}$. Let us stress that we consider single-conclusioned sequents only. This is intended; in what follows we will show why.

---

[8] See Wiśniewski et al. (2005), Wiśniewski and Shangin (2006), Leszczyńska (2007).

2.4 Exemplary formal languages with questions

Clearly, $\mathcal{L}_{cpl}$-sequents are not expressions of $\mathcal{L}_{cpl}$. But (assuming that a sequence of wffs of $\mathcal{L}_{cpl}$ is written down by separating its consecutive elements with commas, as we do here), a $\mathcal{L}_{cpl}$-sequent is an expression of $\mathcal{L}^?_{\vdash cpl}$.

$\mathcal{L}_{cpl}$-sequents perform the role of atomic d-wffs of $\mathcal{L}^?_{\vdash cpl}$. In other words, by an *atomic d-wff* of $\mathcal{L}^?_{\vdash cpl}$ we mean an expression of $\mathcal{L}^?_{\vdash cpl}$ of the form (2.30). We use Greek lower-case letters, ϕ, ψ, with subscripts if needed, as metalanguage variables for atomic d-wffs of $\mathcal{L}^?_{\vdash cpl}$.

Compound d-wffs of $\mathcal{L}^?_{\vdash cpl}$ are built from atomic d-wffs by means of & and/or ng; the construction is standard. Observe that & and ng never occur inside atomic d-wffs. We adopt the usual conventions concerning omitting parentheses in d-wffs of $\mathcal{L}^?_{\vdash cpl}$.

Questions of $\mathcal{L}^?_{\vdash cpl}$ have the form:

$$?(\Phi) \tag{2.31}$$

where Φ is a non-empty and finite sequence of atomic d-wffs of $\mathcal{L}^?_{\vdash cpl}$, that is, of $\mathcal{L}_{cpl}$-sequents. We say that a question of the form (2.31) is *based on* the sequence Φ, and that the terms of this sequence are *constituents* of the question.

Let $\Phi = \phi_1, \ldots, \phi_n$, and let $Q = ?(\Phi)$. The following:

$$(\phi_1 \:\&\: (\phi_2 \:\&\: \ldots \:\&\: (\phi_{n-1} \:\&\: \phi_n)\ldots)) \tag{2.32}$$

$$ng(\phi_1 \:\&\: (\phi_2 \:\&\: \ldots \:\&\: (\phi_{n-1} \:\&\: \phi_n)\ldots)) \tag{2.33}$$

are the *affirmative answer* to Q and the *negative answer* to Q, respectively. The set of *direct answers* to a question of $\mathcal{L}^?_{\vdash cpl}$ is made up of the affirmative answer and the negative answer, exclusively. Thus questions of $\mathcal{L}^?_{\vdash cpl}$ are, in principle, polar questions. The general reading of a question would be: "Is it the case that: ϕ_1 and $\ldots$ and ϕ_n?".

The intuitive meaning of a question of $\mathcal{L}^?_{\vdash cpl}$ can be described as follows. A constituent of a question is a $\mathcal{L}_{cpl}$-sequent. A $\mathcal{L}_{cpl}$-sequent, $S \vdash A$, is CPL-*valid* iff A is true under each CPL-valuation under which all the terms of the sequence S are true; the concept of CPL-valuation is understood in the standard way (see section 3.1.4 of the next chapter). Thus a question asks about *joint validity* of all of its constituents. On the other hand, we may say that a wff A of $\mathcal{L}_{cpl}$ is CPL-entailed by a sequence of wffs S of $\mathcal{L}_{cpl}$ iff the $\mathcal{L}_{cpl}$-sequent $S \vdash A$ is CPL-valid. Moreover, due to the completeness of CPL, the CPL-validity of a $\mathcal{L}_{cpl}$-sequent $S \vdash A$ is tantamount to CPL-derivability of A from (the set of terms of) S. Thus a question of the form:

$$?(S \vdash A) \tag{2.34}$$

asks about CPL-entailment (or CPL-derivability) of A by/from S, and a question of the form:

$$?(\vdash A) \tag{2.35}$$

can read "Is A CPL-valid (is a CPL-thesis)?". Moreover, a question of the form:

$$?(S \vdash p \land \neg p) \tag{2.36}$$

can be regarded as a question about CPL-consistency of (the set made up of the terms of) S. When we have a question based on more than one constituent, it asks about the joint validity of all of its constituents, but gains its specific meaning depending on the forms of these constituents.

3
Elements of Minimal Erotetic Semantics: Declaratives

In this chapter we introduce semantic concepts pertaining to d-wffs, while the next chapter presents some semantic concepts pertaining to e-formulas/questions. In both cases we make use of the conceptual apparatus of *Minimal Erotetic Semantics* (hereafter: MiES).

Generally speaking, MiES enables an introduction of some important semantic notions pertaining to questions regardless of whether – and if so, how – the semantics of questions themselves has been previously elaborated. Moreover, MiES provides a uniform framework for dealing with semantics of "declarative parts" of formal languages with questions. As for the "erotetic parts", one makes use of already given assignments of direct answers to questions/e-formulas, and of semantic concepts pertaining to declaratives.

MiES combines some ideas present in Belnap's erotetic semantics (cf. Belnap and Steel (1976)) with certain insights to be found in the book Shoesmith and Smiley (1978). Of course, MiES also goes beyond them.

3.1 Admissible partitions and entailment

Let $\mathcal{L}$ be a formal language in which both d-wffs and e-formulas/questions occur. First, we make use of the concept of partition of a language. We follow here the idea of Shoesmith and Smiley (1978), adjusting it a little bit for our purposes.

Let $\mathcal{D}_\mathcal{L}$ designate the set of d-wffs of $\mathcal{L}$.

Definition 3.1 (*Partition of the set of d-wffs*). *A partition of $\mathcal{D}_\mathcal{L}$ is an ordered pair:*

$$\mathsf{P} = \langle \mathsf{T}_\mathsf{P}, \mathsf{U}_\mathsf{P} \rangle$$

where $\mathsf{T}_\mathsf{P} \cap \mathsf{U}_\mathsf{P} = \emptyset$ *and* $\mathsf{T}_\mathsf{P} \cup \mathsf{U}_\mathsf{P} = \mathcal{D}_\mathcal{L}$.

Intuitively, T_P consists of all the d-wffs which are "true" in P, and U_P is made up of all the d-wffs which are "untrue" in P. But "true" is used here as a cover term which, as we will see, is construed differently in different cases and is not synonymous with "true in the actual world". For brevity, however, it is convenient to speak about truths and untruths of a partition.

Definition 3.2 (*Partition of a language*). *A partition of $\mathcal{L}$ is a partition of $\mathcal{D}_\mathcal{L}$.*

Note that we have used the term "partition" as pertaining to the set of d-wffs only. What is "partitioned" is neither the "logical space" nor the set of e-formulas.

When the sets of questions and d-wffs are disjoint, a question is neither in $\mathsf{T_P}$ nor in $\mathsf{U_P}$, for any partition P. In general, MiES does not presuppose that questions are true or false.

The concept of partition introduced above is very general and admits partitions which are rather odd from the intuitive point of view. For example, there are partitions in which $\mathsf{T_P}$ is a singleton set, or in which $\mathsf{U_P}$ is the empty set. In order to avoid oddity on the one hand, and to reflect some basic facts about the language just considered on the other, we should distinguish a class of *admissible partitions*, being a non-empty subclass of the class of all partitions of the language. This step allows us to define a series of useful semantic concepts, in particular the concept of entailment.

3.1.1 Entailment

Let X stand for a set of d-wffs of a language $\mathcal{L}$ of the considered kind, and let A be a d-wff of $\mathcal{L}$. Entailment in $\mathcal{L}$, symbolized by $\models_\mathcal{L}$, is defined as follows ($\subset$ is the sign of inclusion):

Definition 3.3 (*Entailment*). $X \models_\mathcal{L} A$ *iff there is no admissible partition* $\mathsf{P} = \langle \mathsf{T_P}, \mathsf{U_P} \rangle$ *of $\mathcal{L}$ such that $X \subset \mathsf{T_P}$ and $A \in \mathsf{U_P}$.*

What remains to be done is to characterize the class of admissible partitions. There are (at least) three methods of doing it: the minimalistic one, the direct one, and the indirect one.

3.1.2 The minimalistic method

Let us consider a language $\mathcal{L}$ of the analysed kind such that the declarative part of $\mathcal{L}$ is the language of a logic ℓ; the d-wffs of $\mathcal{L}$ are just the wffs of the language of ℓ. A logic determines the corresponding consequence relation; it is a binary relation between sets of wffs on the one hand and individual formulas on the other. Let $\vdash_\ell$ stand for the consequence relation determined by ℓ. We assume that $\vdash_\ell$ satisfies the following standard conditions (see Shoesmith and Smiley (1978)):

(OVERLAP) If $A \in X$, then $X \vdash_\ell A$,
(DILUTION) If $X \vdash_\ell A$ and $X \subset Y$, then $Y \vdash_\ell A$,
(CUT FOR SETS) If $X \cup Y \vdash_\ell A$ and $X \vdash_\ell B$ for every $B \in Y$, then $X \vdash_\ell A$,

for any sets X, Y of d-wffs of the language of ℓ, and any d-wffs A, B of the language.

Assume that $\vdash_\ell$ is not universal. We say that a partition $\mathsf{P} = \langle \mathsf{T_P}, \mathsf{U_P} \rangle$ of $\mathcal{L}$ is *improper* iff for some set X of d-wffs of $\mathcal{L}$ and some d-wff A of $\mathcal{L}$ such that $X \vdash_\ell A$ we have: $X \subset \mathsf{T_P}$ and $A \in \mathsf{U_P}$; otherwise P is called *proper*.

Now we have two options. First, the class of admissible partitions of $\mathcal{L}$ could have been identified with the class of all proper partitions of the language. As a consequence we would get:

Corollary 3.4. $\vdash_\ell \,\subset\, \models_\mathcal{L}$.

Corollary 3.4 yields that all the inferences licensed by ℓ are $\mathcal{L}$-valid, that is, their conclusions are entailed in $\mathcal{L}$ by the premises. Thus when it comes to applications, one can make use of the "full strength" of ℓ. However, it is possible that there are $\mathcal{L}$-valid inferences which are not licensed by ℓ and, what is worse, the construction does not determine them.

The second option is to define the class of admissible partitions of $\mathcal{L}$ as the class of proper partitions of the language that fulfils the following condition:

($\heartsuit$) if $X \nvdash_\ell A$, then for some partition $\mathsf{P} = \langle \mathsf{T_P}, \mathsf{U_P} \rangle$ in the class:
$X \subset \mathsf{T_P}$ and $A \in \mathsf{U_P}$

for any set of d-wffs X of $\mathcal{L}$ and any d-wff A of the language. We get:

Corollary 3.5. $\vdash_\ell \,=\, \models_\mathcal{L}$.

Now entailment in $\mathcal{L}$ amounts, set-theoretically, to the consequence relation determined by ℓ. This facilitates possible applications. Let us stress that the above construction permits that ℓ is a non-classical logic (but, still, a monotonic logic).

Comments. The general framework of MiES allows both Classical Logic and a non-classical logic to serve as the underlying logic of d-wffs. The virtue of the minimalistic method is its generality (although the method is not applicable in all cases). However, the relevant concept of truth (of a d-wff in an admissible partition) is, as a matter of fact, left undetermined conceptually.

3.1.3 The direct method: language $\mathcal{L}^?_{\vdash cpl}$

Atomic d-wffs of language $\mathcal{L}^?_{\vdash cpl}$ (cf. section 2.4.4 of Chapter 2) involve the turnstile symbol $\vdash$, intuitively interpreted as referring to CPL-entailment/derivability. $\mathcal{L}^?_{\vdash cpl}$ is still an object-level language, however. The class of admissible partitions of $\mathcal{L}^?_{\vdash cpl}$ is defined according to the direct method, by imposing some conditions on partitions of $\mathcal{L}^?_{\vdash cpl}$. These conditions reflect certain basic properties of CPL-entailment/derivability.

As a preparatory step, we distinguish α-wffs and β-wffs of the initial language $\mathcal{L}_{cpl}$ of CPL. The distinction originates from Smullyan (1968). α-wffs are of the form:

$$A \wedge B \tag{3.1}$$
$$\neg(A \vee B) \tag{3.2}$$
$$\neg(A \to B) \tag{3.3}$$

whereas β-wffs have the forms:

$$\neg(A \wedge B) \tag{3.4}$$
$$A \vee B \tag{3.5}$$
$$A \to B \tag{3.6}$$

Contrary to Smullyan, we do not consider $\neg\neg A$ as an α-wff. To each α-wff we assign two wffs: α_1, α_2, and to each β-wff we assign three wffs: $\beta_1, \beta_2, \beta_1^*$. The assignment is given by Table 3.1.

Table 3.1. α/β formulas.

α	α_1	α_2	β	β_1	β_2	β_1^*
$A \wedge B$	A	B	$\neg(A \wedge B)$	$\neg A$	$\neg B$	A
$\neg(A \vee B)$	$\neg A$	$\neg B$	$A \vee B$	A	B	$\neg A$
$\neg(A \to B)$	A	$\neg B$	$A \to B$	$\neg A$	B	A

β_1^* may be called the *complement* of β_1.

Recall that atomic d-wffs of $\mathcal{L}^?_{\vdash cpl}$ are of the form $S \vdash A$, where A is a wff of $\mathcal{L}_{cpl}$ and S is a sequence of wffs of $\mathcal{L}_{cpl}$. The turnstile $\vdash$ is supposed to represent CPL-entailment/derivability. It is well-known that an α-wff is CPL-entailed by a set of CPL-wffs X if, and only if both α_1 and α_2 are CPL-entailed by X, and that a β-wff is CPL-entailed by X if, and only if β_2 is CPL-entailed by X enriched with β_1^*. Similarly, CPL-entailment from premises which involve an α-wff is tantamount to CPL-entailment from the corresponding premises which involve, instead of α, both α_1 and α_2. When we have a β-wff among the premises, a wff is CPL-entailed by these premises just in case the wff is CPL-entailed both by the premises which involve β_1 instead of β and the premises that involve β_2 instead of β. Finally, CPL-entailment of/from $\neg\neg A$ amounts to CPL-entailment of/from A. We make use of these observations when defining admissible partitions of language $\mathcal{L}^?_{\vdash cpl}$.

In what follows the symbol $'$ stands for the concatenation-sign for sequences of wffs of $\mathcal{L}_{cpl}$. Thus $S \, ' \, T$ is the concatenation of a sequence of wffs S and a sequence of wffs T. An expression of the form $S \, ' \, A$ represents the concatenation of S and the one-term sequence whose term is A. Of course, $S \, ' \, A \, ' \, T$ is the concatenation of $S \, ' \, A$ and T. We use $\mathfrak{r}, \mathfrak{u}$ as metalanguage variables for d-wffs of $\mathcal{L}^?_{\vdash cpl}$.

Definition 3.6 (*Admissible partitions of* $\mathcal{L}^?_{\vdash cpl}$). *A partition* $\mathsf{P} = \langle \mathsf{T_P}, \mathsf{U_P} \rangle$ *of* $\mathcal{L}^?_{\vdash cpl}$ *is admissible iff the following conditions hold:*

1. $\ulcorner S \vdash \alpha \urcorner \in \mathsf{T_P}$ *iff* $\ulcorner S \vdash \alpha_1 \urcorner \in \mathsf{T_P}$ *and* $\ulcorner S \vdash \alpha_2 \urcorner \in \mathsf{T_P}$;
2. $\ulcorner S \, ' \, T \vdash \beta \urcorner \in \mathsf{T_P}$ *iff* $\ulcorner S \, ' \, \beta_1^* \, ' \, T \vdash \beta_2 \urcorner \in \mathsf{T_P}$;

3.1 Admissible partitions and entailment

3. $\ulcorner S\,'\alpha\,' T \vdash C \urcorner \in \mathsf{T_P}$ iff $\ulcorner S\,'\alpha_1\,'\alpha_2\,' T \vdash C \urcorner \in \mathsf{T_P}$;
4. $\ulcorner S\,'\beta\,' T \vdash C \urcorner \in \mathsf{T_P}$ iff $\ulcorner S\,'\beta_1\,' T \vdash C \urcorner \in \mathsf{T_P}$ and
   $\ulcorner S\,'\beta_2\,' T \vdash C \urcorner \in \mathsf{T_P}$;
5. $\ulcorner S \vdash \neg\neg A \urcorner \in \mathsf{T_P}$ iff $\ulcorner S \vdash A \urcorner \in \mathsf{T_P}$;
6. $\ulcorner S\,'\neg\neg A\,' T \vdash B \urcorner \in \mathsf{T_P}$ iff $\ulcorner S\,'A\,' T \vdash B \urcorner \in \mathsf{T_P}$;
7. $\ulcorner \mathfrak{r}\,\&\,\mathfrak{u} \urcorner \in \mathsf{T_P}$ iff $\mathfrak{r} \in \mathsf{T_P}$ and $\mathfrak{u} \in \mathsf{T_P}$;
8. if $\mathfrak{u} \notin \mathsf{T_P}$, then $\ulcorner ng\,\mathfrak{u} \urcorner \in \mathsf{T_P}$;
9. if $\mathfrak{u} \in \mathsf{T_P}$, then $\ulcorner ng\,\mathfrak{u} \urcorner \notin \mathsf{T_P}$.

Entailment in $\mathcal{L}^?_{\vdash cpl}$ can be defined according to the schema presented by Definition 3.3. In particular, a d-wff $\mathfrak{u}$ is entailed by a d-wff $\mathfrak{r}$ iff there is no admissible partition $\mathsf{P} = \langle \mathsf{T_P}, \mathsf{U_P} \rangle$ of $\mathcal{L}^?_{\vdash cpl}$ such that $\mathfrak{r} \in \mathsf{T_P}$ and $\mathfrak{u} \in \mathsf{U_P}$.

Let us stress that entailment in $\mathcal{L}^?_{\vdash cpl}$ should not be confused with the relation represented by $\vdash$. Entailment in $\mathcal{L}^?_{\vdash cpl}$ is a relation between d-wffs of $\mathcal{L}^?_{\vdash cpl}$, defined in the metalanguage of $\mathcal{L}^?_{\vdash cpl}$. The turnstile, $\vdash$, is the sign of the object-level language $\mathcal{L}^?_{\vdash cpl}$ and refers to entailment/derivability in the initial language $\mathcal{L}_{cpl}$ of CPL. So a statement of the form:

$$\ulcorner S \vdash A \urcorner \models_{\mathcal{L}^?_{\vdash cpl}} \ulcorner T \vdash B \urcorner$$

claims that CPL-entailment/derivability among S and A yields CPL-entailment /derivability between T and B, and similarly in more complex cases.

Consider the following questions:

$$?(p \to q \vdash \neg q \to \neg p) \qquad (3.7)$$

$$?(\neg p, \neg q \vdash \neg p\,;\,q, \neg q \vdash \neg p) \qquad (3.8)$$

The affirmative answers to the above questions entail each other in $\mathcal{L}^?_{\vdash cpl}$, and similarly for the negative answers. Thus (3.7) and (3.8) are, in a sense, equivalent. Hence the problem of CPL-validity of:

$$p \to q \vdash \neg q \to \neg p$$

reduces to the problem of *joint* CPL-validity of sequent $\neg p, \neg q \vdash \neg p$ and sequent $q, \neg q \vdash \neg p$. We will come back to this issue in Chapter 8.

3.1.4 The indirect method: language $\mathcal{L}^?_{cpl}$

Generally speaking, the indirect method applies a full-fledged semantics of the declarative part of a language as the basis.

Language $\mathcal{L}^?_{cpl}$ results from the language $\mathcal{L}_{cpl}$ of CPL by enriching $\mathcal{L}_{cpl}$ with questions (see section 2.4.1 of Chapter 2 for details). The language of CPL, however, has well-defined semantics. We take the standard one and we introduce, first, the concept of CPL-valuation.

Let **1** and **0** stand for Truth and Falsehood, respectively. Recall that the set $\mathcal{D}_{\mathcal{L}^?_{cpl}}$ is the set of wffs of $\mathcal{L}_{cpl}$ (i.e. of CPL-formulas). A CPL-*valuation* is a function $v : \mathcal{D}_{\mathcal{L}^?_{cpl}} \mapsto \{\mathbf{1}, \mathbf{0}\}$ such that:

- for each propositional variable p, either $v(\mathbf{p}) = \mathbf{1}$ or $v(\mathbf{p}) = \mathbf{0}$;
- $v(\neg A) = \mathbf{1}$ iff $v(A) = \mathbf{0}$;
- $v(A \wedge B) = \mathbf{1}$ iff $v(A) = v(B) = \mathbf{1}$;
- $v(A \vee B) = \mathbf{1}$ iff $v(A) = \mathbf{1}$ or $v(B) = \mathbf{1}$;
- $v(A \to B) = \mathbf{1}$ iff $v(A) = \mathbf{0}$ or $v(B) = \mathbf{1}$.

Definition 3.7 (*Admissible partitions of $\mathcal{L}^?_{cpl}$*). *A partition* $\mathsf{P} = \langle \mathsf{T_P}, \mathsf{U_P} \rangle$ *of $\mathcal{L}^?_{cpl}$ is admissible iff for some CPL-valuation v:*

- $\mathsf{T_P} = \{A \in \mathcal{D}_{\mathcal{L}^?_{cpl}} : v(A) = \mathbf{1}\}$, *and*
- $\mathsf{U_P} = \{B \in \mathcal{D}_{\mathcal{L}^?_{cpl}} : v(B) = \mathbf{0}\}$.

Thus the set of "truths" of an admissible partition equals the set of d-wffs which are true under the corresponding CPL-valuation.

Observe that all the usual semantic properties are retained, but now they can be rephrased in terms of admissible partitions. For example, there is no admissible partition P of $\mathcal{L}^?_{cpl}$ such that both A and $\neg A$ belong to $\mathsf{T_P}$. Moreover, $A \wedge B$ belongs to $\mathsf{T_P}$ if, and only if A is in $\mathsf{T_P}$ and B is in $\mathsf{T_P}$, and analogous classical clauses hold for other connectives.

Note finally that entailment in $\mathcal{L}^?_{cpl}$ reduces to CPL-entailment. The following is true:

Corollary 3.8. $X \models_{\mathcal{L}^?_{cpl}} A$ *iff there is no CPL-valuation v such that $v(B) = \mathbf{1}$ for all $B \in X$, and $v(A) = \mathbf{0}$.*

3.1.5 The indirect method: language $\mathcal{L}^?_{S4}$

Let us now turn to modal propositional languages with questions characterized in section 2.4.2 of Chapter 2. As an illustration, we consider the case of the (propositional) modal logic S4 and the corresponding language $\mathcal{L}^?_{S4}$.

The set $\mathcal{D}_{\mathcal{L}^?_{S4}}$ of d-wffs of $\mathcal{L}^?_{S4}$ equals the set $\mathcal{D}_{\mathcal{L}_{S4}}$ of wffs of the language $\mathcal{L}_{S4}$ of S4.

We start by using a semantics of S4. Since the logic has a standard relational semantics, we apply it to the d-wffs of $\mathcal{L}^?_{S4}$. However, we are interested in truth in a world rather than in truth in a model.

A S4-model is an ordered triple:

$$\langle W, R, V \rangle \tag{3.9}$$

where $W \neq \emptyset$, $R \subset W \times W$ is both reflexive and transitive in W, and $V : \mathcal{D}_{\mathcal{L}_{S4}} \times W \mapsto \{\mathbf{1}, \mathbf{0}\}$ satisfies the following conditions, for any $w \in W$:

- for each propositional variable p, either $V(\mathbf{p}, w) = \mathbf{1}$ or $V(\mathbf{p}, w) = \mathbf{0}$;
- $V(\neg A, w) = \mathbf{1}$ iff $V(A, w) = \mathbf{0}$;
- $V(A \wedge B, w) = \mathbf{1}$ iff $V(A, w) = V(B, w) = \mathbf{1}$;
- $V(A \vee B, w) = \mathbf{1}$ iff $V(A, w) = \mathbf{1}$ or $V(B, w) = \mathbf{1}$;
- $V(A \to B, w) = \mathbf{1}$ iff $V(A, w) = \mathbf{0}$ or $V(B, w) = \mathbf{1}$;

- $V(\Diamond A, w) = 1$ iff for some $w^* \in R^{\rightarrow}w : V(A, w^*) = 1$;
- $V(\Box A, w) = 1$ iff for each $w^* \in R^{\rightarrow}w : V(A, w^*) = 1$.

As usual, elements of W are called (possible) worlds, and R is the accessibility relation. Let $\boldsymbol{M} = \langle W, R, V \rangle$ be a S4-model. *Truth* of a d-wff A in a world w of model $\boldsymbol{M}$, in symbols $(\boldsymbol{M}, w) \models A$, is defined by:

$$(\boldsymbol{M}, w) \models A \text{ iff } V(A, w) = 1 \qquad (3.10)$$

Admissible partitions of $\mathcal{L}_{S4}^?$ are defined as follows:

Definition 3.9 (Admissible partitions of $\mathcal{L}_{S4}^?$). *A partition* $\mathsf{P} = \langle \mathsf{T_P}, \mathsf{U_P} \rangle$ *of $\mathcal{L}_{S4}^?$ is admissible iff for some S4-model $\boldsymbol{M} = \langle W, R, V \rangle$ and for some $w \in W$:*

- $\mathsf{T_P} = \{A \in \mathcal{D}_{\mathcal{L}_{S4}^?} : (\boldsymbol{M}, w) \models A\}$, and
- $\mathsf{U_P} = \{B \in \mathcal{D}_{\mathcal{L}_{S4}^?} : (\boldsymbol{M}, w) \not\models B\}$.

Thus the set of "truths" of an admissible partition consists of all the d-wffs which are true in the corresponding world of a given model.

Note that the theses of S4 are always included in $\mathsf{T_P}$, for any admissible partition P of the language. This is how it should be. Moreover, the following holds:

Corollary 3.10.
$X \models_{\mathcal{L}_{S4}^?} A$ iff there is no S4-model $\boldsymbol{M} = \langle W, R, V \rangle$ such that for some $w \in W$: $(\boldsymbol{M}, w) \models B$ for each $B \in X$, and $(\boldsymbol{M}, w) \not\models A$.

Thus entailment in $\mathcal{L}_{S4}^?$ reduces to the so-called local entailment in S4.[1]

We have considered above the case of S4. As long as other normal propositional modal logics are concerned, one can proceed analogously; the only difference lies in conditions to be imposed on the accessibility relation. When non-normal modal logics constitute the background, the concept of model (usually) gets more complicated.

Remark. The above construction relies on the assumption that the declarative part of a language is the language of a fixed modal logic. On the other hand, it happens that languages of distinct modal logics are syntactically indistinguishable. However, they are still distinct languages, since the meanings of modalities depend on the underlying logics. This should be, and in fact is reflected at the level of MiES.

3.1.6 The indirect method: language $\mathcal{L}_{fom}^?$

The declarative part of $\mathcal{L}_{fom}^?$ is the language of Monadic First-Order Logic with Identity (and no function symbols). We use the model-theoretical semantics as the point of departure. By a *model* of $\mathcal{L}_{fom}^?$ we mean an ordered pair $\langle M, f \rangle$ such that M is a non-empty set, and f is a function which assigns an element

---

[1] If, for some reasons, one would need reduction to global entailment, admissible partitions are to be defined in terms of truth in a model.

of M to each individual constant of $\mathcal{L}^?_{fom}$, and a subset of M to each unary predicate of $\mathcal{L}^?_{fom}$.

Let $\mathcal{M} = \langle M, f \rangle$ be a model of $\mathcal{L}^?_{fom}$. A $\mathcal{M}$-*valuation* is a denumerable sequence of elements of M. The concepts of value of a term under a $\mathcal{M}$-valuation, and of satisfaction of a d-wff by a $\mathcal{M}$-valuation are defined in the standard manner. A wff A is *true* in a model $\mathcal{M} = \langle M, f \rangle$, in symbols $\mathcal{M} \models A$, iff A is satisfied by each $\mathcal{M}$-valuation.

In the second step we define the class of normal models. Roughly, a model $\mathcal{M} = \langle M, f \rangle$ of $\mathcal{L}^?_{fom}$ is *normal* just in case all the elements of M are named by individual constants of $\mathcal{L}^?_{fom}$. To be more precise, by a normal model of $\mathcal{L}^?_{fom}$ we mean a model $\mathcal{M} = \langle M, f \rangle$ of the language such that for each $y \in M$ we have: $y = f(c_i)$ for some individual constant c_i of $\mathcal{L}^?_{fom}$.

As long as normal models are concerned, the truth of an existential generalization $\exists x A x$ warrants the existence of a true direct answer to the corresponding existential which-question $?S(Ax)$. This is why we have distinguished these models here.

Definition 3.11 (*Admissible partitions of $\mathcal{L}^?_{fom}$*). *A partition* $\mathsf{P} = \langle \mathsf{T_P}, \mathsf{U_P} \rangle$ *of $\mathcal{L}^?_{fom}$ is admissible iff for some normal model $\mathcal{M} = \langle M, f \rangle$ of $\mathcal{L}^?_{fom}$:*

- $\mathsf{T_P} = \{A \in \mathcal{D}_{\mathcal{L}^?_{fom}} : \mathcal{M} \models A\}$, *and*
- $\mathsf{U_P} = \{B \in \mathcal{D}_{\mathcal{L}^?_{fom}} : \mathcal{M} \not\models B\}$.

Hence the set of "truths" of an admissible partition equals the set of d-wffs which are true in the corresponding normal model.

One can prove the following:[2]

Corollary 3.12. *Let X be a finite set of d-wffs of $\mathcal{L}^?_{fom}$. $X \models_{\mathcal{L}^?_{fom}} A$ iff A is true in each model of $\mathcal{L}^?_{fom}$ in which all the d-wffs in X are true.*

Thus in the case of finite sets of d-wffs, entailment in $\mathcal{L}^?_{fom}$ reduces to logical entailment. The situation changes, however, when infinite sets of d-wffs are taken into consideration.

Remark. The reference to normal models is the key feature of the above construction. We have distinguished them for "erotetic" reasons. However, when we deal with a first-order (or a higher-order) language enriched with questions, normal models can be distinguished for many reasons and in different manners. For example, one can define them as models of a theory expressed in the declarative part of the language[3], or as models which make true some definition(s). It is also permitted to consider all models as normal. Each decision determines the corresponding entailment relation.

---

[2] For a proof, see e.g. Wiśniewski (1995), p. 125.
[3] That is, models of the language in which all the theorems are true.

3.2 Multiple-conclusion entailment

It is natural to think of questions which have well-defined sets of direct answers as offering sets of "possibilities" or "alternatives", among which some selection or choice is requested to be made. And when we are going to analyse relations between questions and contexts of their appearance, some notion of, to speak generally, "entailing a set of possibilities" is needed. There is a logic, however, within which such notion is elaborated on: it is multiple-conclusion logic (see Shoesmith and Smiley (1978)). This logic generalizes the concept of entailment, regarding it as a relation between sets of declarative formulas (d-wffs). The entailed set is conceived as, intuitively speaking, setting out the field within which the truth must lie if the premises are all true.

Let $\mathcal{L}$ be a language of the kind considered above, and let X and Y be sets of d-wffs of $\mathcal{L}$. The relation $\models_{\mathcal{L}}$ of *multiple-conclusion entailment in* $\mathcal{L}$ is defined as follows:

Definition 3.13 (*Multiple-conclusion entailment*). $X \models_{\mathcal{L}} Y$ *iff there is no admissible partition* $\mathsf{P} = \langle \mathsf{T_P}, \mathsf{U_P} \rangle$ *of* $\mathcal{L}$ *such that* $X \subset \mathsf{T_P}$ *and* $Y \subset \mathsf{U_P}$.

Thus X multiple-conclusion entails (*mc-entails* for short) Y if, and only if there is no admissible partition in which X consists of truths and Y consists of untruths. In other words, mc-entailment between X and Y holds just in case the truth of all the d-wffs in X warrants the presence of some true d-wff(s) in Y: whenever all the d-wffs in X are true in an admissible partition P, then at least one d-wffs in Y is true in the partition P.

Here are simple examples:

$$\{p \vee q\} \models_{\mathcal{L}^?_{cpl}} \{p, q\} \tag{3.11}$$

$$\{p \wedge q \to r, \neg r\} \models_{\mathcal{L}^?_{cpl}} \{\neg p, \neg q\} \tag{3.12}$$

$$\{\exists \mathsf{xPx}\} \models_{\mathcal{L}^?_{fom}} \mathsf{S}(\mathsf{Px}) \tag{3.13}$$

where P is an unary predicate of $\mathcal{L}^?_{fom}$;[4]

$$\{\neg \forall \mathsf{xPx}\} \models_{\mathcal{L}^?_{fom}} \mathsf{S}(\neg \mathsf{Px}) \tag{3.14}$$

$$\{\exists \mathsf{xPx}, \neg \mathsf{Pa}\} \models_{\mathcal{L}^?_{fom}} \mathsf{S}(\mathsf{Px}) \setminus \{\mathsf{Pa}\} \tag{3.15}$$

where a is an individual constant.

$$\{\Diamond p, \Diamond \neg p\} \models_{\mathcal{L}^?_{S4}} \{p, \neg p\} \tag{3.16}$$

$$\{\Box(p \to q), \Diamond \neg q\} \models_{\mathcal{L}^?_{S4}} \{\Diamond q, \Box \neg p\} \tag{3.17}$$

$$\{\neg \Box p\} \models_{\mathcal{L}^?_{S4}} \{\Box \neg \Box p, \Diamond \Box p\} \tag{3.18}$$

---

[4] Recall that $\mathsf{S}(\mathsf{Px})$ is the set of sentences of $\mathcal{L}^?_{fom}$ of the form Pc, where c stands for an individual constant.

3.2.1 Multiple-conclusion entailment vs. single-conclusion entailment

Definition 3.13 offers a non-trivial generalization of the concept of entailment. Observe that in neither of the above cases (3.11) – (3.18) a single element of the mc-entailed set is entailed, in the sense of Definition 3.3, by the premises. So mc-entailment cannot be defined in terms of (single-conclusion) entailment only. Intuitively speaking, mc-entailment between X and Y amounts to (single-conclusion) entailment of a "disjunction" of all the Y's from a "conjunction" of all the X's. However, we cannot use this idea as a basis for a definition of mc-entailment. There are languages of the considered kind in which disjunction does not occur (for example, language $\mathcal{L}^?_{\vdash cpl}$) or is understood differently than in Classical Logic. Moreover, Y can be an infinite set and an infinite "disjunction" need not be expressible in a language (similarly for X and conjunction).

Single-conclusion entailment, however, can be conceived as a special case of mc-entailment. The following holds:

Corollary 3.14. $X \models_{\mathcal{L}} A$ iff $X \models\models_{\mathcal{L}} \{A\}$.

Thus one can define entailment as mc-entailment of a singleton set.

For conciseness, we will write $A \models\models_{\mathcal{L}} Y$ instead of $\{A\} \models\models_{\mathcal{L}} Y$, and $X \models\models_{\mathcal{L}} A$ instead of $X \models\models_{\mathcal{L}} \{A\}$. If $\{A\} \models\models_{\mathcal{L}} Y$, we also say that Y is mc-entailed by the d-wff A.

The concept of mc-entailment proved its usefulness in the logic of questions in many ways.[5] It will serve as one of the main conceptual tools used in the next sections of this chapter, and in consecutive chapters.

3.3 Eliminating and narrowing down

Although classical negation occurs in any of the languages considered above, this is not a general rule. For this reason we need a technical (semantic) concept of elimination. We put:

Definition 3.15 (Elimination). *Let B, C be d-wffs of $\mathcal{L}$, and let X, Y be sets of d-wffs of $\mathcal{L}$.*

1. *X eliminates Y iff for each admissible partition $\mathsf{P} = \langle \mathsf{T_P}, \mathsf{U_P} \rangle$ of $\mathcal{L}$:*
   (∘) *if $X \subset \mathsf{T_P}$, then $Y \subset \mathsf{U_P}$;*
2. *X eliminates C iff X eliminates $\{C\}$;*
3. *B eliminates Y iff $\{B\}$ eliminates Y;*
4. *B eliminates C iff $\{B\}$ eliminates $\{C\}$.*

Thus C is eliminated by B if C is false in each admissible partition in which B is true, and similarly in the remaining cases.

For single d-wffs elimination is symmetric. One can easily prove:

Corollary 3.16. *If B eliminates C, then C eliminates B.*

---
[5] Starting from Buszkowski (1989) as well as Wiśniewski (1989).

3.3 Eliminating and narrowing down

Clearly eliminating is weaker than contradicting: it happens that B eliminates C, and both B and C are false in an admissible partition. For example, take language $\mathcal{L}_{cpl}^?$ and the following d-wffs of the language:

$$p \wedge q \tag{3.19}$$

$$p \wedge \neg q \tag{3.20}$$

(3.19) eliminates (3.20), but (3.19) and (3.20) are false in any (admissible) partition of $\mathcal{L}_{cpl}^?$ in which p is false.

Definition 3.17 (Narrowing down). *A set of d-wffs X of $\mathcal{L}$ narrows down a set of d-wffs Y of $\mathcal{L}$ iff there exists a non-empty proper subset Y^* of Y such that $X \models_{\mathcal{L}} Y^*$.*

The underlying intuition is: a disjunction of some, but not all elements of Y (that is, a disjunction whose disjuncts do not exhaust the whole Y) is entailed by X. The disjunction can be infinite, but "classical" in the sense that it is true in an admissible partition just in case at least one disjunct is true in the partition.

Narrowing down is not tantamount to elimination. For instance, let $X = \{p, p \to q \vee r\}$ and $Y = \{q, r, t\}$. Clearly, X mc-entails the proper subset $\{q, r\}$ of Y, but does not eliminate any d-wff in Y. Similarly, elimination need not yield narrowing down. For example, $\neg q$ eliminates the d-wff q of Y, but does not mc-entail any proper subset of Y.

4

Elements of Minimal Erotetic Semantics: Questions

In this chapter we introduce some semantic concepts pertaining to questions. We do it within the MiES framework.

From now on, we will be considering (unless otherwise stated) an arbitrary but fixed formal language $\mathcal{L}$ of the analysed kind; by d-wffs and questions we will mean d-wffs and questions of the language. Language $\mathcal{L}$ is supposed to satisfy the following conditions: (*a*) it has questions and d-wffs among well-formed expressions, (*b*) for any question of the language, the set of direct answers to the question is defined, (*c*) direct answers are d-wffs, and (*d*) the class of admissible partitions of $\mathcal{L}$ is defined. For the sake of brevity, in what follows we omit the specifications "of $\mathcal{L}$" and "in $\mathcal{L}$". Similarly, we write $\models$ instead of $\models_\mathcal{L}$, and $\models$ instead of $\models_\mathcal{L}$.

We use $\mathsf{d}Q$ for the set of all the direct answers to question Q.

We say that a d-wff A is *true in a partition* $\mathsf{P} = \langle \mathsf{T}_\mathsf{P}, \mathsf{U}_\mathsf{P} \rangle$ if $A \in \mathsf{T}_\mathsf{P}$, and *false in* P if $A \in \mathsf{U}_\mathsf{P}$.

4.1 Soundness of a question

MiES does not presuppose that questions are true or false. Instead, the concept of soundness of a question is used.

The underlying intuition is: a question Q is sound if, and only if at least one direct answer to Q is true.[1] So, for example, the question "Who was the only wife of Henry the Eight?" is not sound, whereas "Who was the first wife of Henry the Eight?" is sound. Similarly, the question "For which $x \in \mathbf{N}$, $x = 2 \div 0$?" is not sound, while "For which $x \in \mathbf{N}$, $x = 2 \div 1$?" is sound.

Of course, when a formal language is concerned, the concept of soundness needs a relativization. So we put:

Definition 4.1 (*Soundness*). *A question Q is sound in a partition P iff $\mathsf{d}Q \cap \mathsf{T}_\mathsf{P} \neq \emptyset$.*

---

[1] The basic idea of this definition is due to Bromberger; see Bromberger (1992), p. 146.

Thus a question is sound in a partition iff at least one direct answer to the question is true in the partition. In practice, we relativize the concept of soundness even further: we are interested in soundness in an admissible partition.

4.2 Safety and riskiness

It may happen that a question is sound in one admissible partition and is not sound in some other(s). If, however, a question is sound in each admissible partition of a language, we call it a safe question.[2] More formally:

Definition 4.2 (Safety). *A question Q is safe iff $\mathsf{d}Q \cap \mathsf{T_P} \neq \emptyset$ for each admissible partition P.*

Observe that a question can be safe although no direct answer to it is valid, that is, true in each admissible partition of a language. For example, the following questions of $\mathcal{L}^?_{cpl}$ are safe, but no direct answer to them is valid:

$$?p \tag{4.1}$$

$$? \pm |p, q| \tag{4.2}$$

However, we still have:[3]

Corollary 4.3. *A question Q is safe iff $\emptyset \models \mathsf{d}Q$.*

A question which is not safe is called risky.[4] To be more precise:

Definition 4.4 (Riskiness). *A question Q is risky iff $\mathsf{d}Q \cap \mathsf{T_P} = \emptyset$ for some admissible partition P.*

Thus a risky question is a question which has no true direct answer in at least one admissible partition of the language. Here are simple examples of risky questions of the language $\mathcal{L}^?_{cpl}$:

$$?\{p, q\} \tag{4.3}$$

$$?\{p \wedge q, p \wedge \neg q\} \tag{4.4}$$

The following questions of $\mathcal{L}^?_{S4}$ are risky:

$$?\{\Box p, \Box \neg p\} \tag{4.5}$$

$$?\{\Box \Diamond p, \Diamond \Box p\} \tag{4.6}$$

Definition 4.5 (Contingency). *A question Q is contingent iff there exist admissible partitions P, P^* such that $\mathsf{d}Q \cap \mathsf{T_P} \neq \emptyset$ and $\mathsf{d}Q \cap \mathsf{T_{P^*}} = \emptyset$.*

---

[2] This idea, in turn, comes from Belnap. See e.g. Belnap and Steel (1976), p. 130.
[3] Recall that Y can be mc-entailed by X although no element of Y is entailed by X; cf. section 3.2.1 of Chapter 3.
[4] Again, this is Belnap's term.

4.3 Presuppositions and prospective presuppositions

Thus contingent questions form a sub-class of risky questions: a question is contingent just in case it is sound in some, but not all admissible partitions. For instance, questions (4.3), (4.4), (4.5) and (4.6) are contingent. The following question of $\mathcal{L}^?_{cpl}$:

$$?\{p \wedge \neg p, \neg(p \vee \neg p)\} \tag{4.7}$$

is not contingent.

Corollary 4.6. *A question Q is contingent iff $\emptyset \not\Vdash \mathsf{d}Q$, but $\emptyset$ does not eliminate $\mathsf{d}Q$.*

A language of the considered kind usually involves both safe and risky questions. There are exceptions, however. For instance, each question of the language $\mathcal{L}^?_{\vdash cpl}$ is safe. This is due to clause (8) of Definition 3.6 of admissible partitions of $\mathcal{L}^?_{\vdash cpl}$.

4.3 Presuppositions and prospective presuppositions

Following Belnap's proposal[5], we define the concept of a presupposition of a question by:

Definition 4.7 (Presupposition). *A d-wff B is a presupposition of a question Q iff $A \models B$ for each $A \in \mathsf{d}Q$.*

Thus a presupposition of a question is a d-wff which is entailed by each direct answer to the question. For instance, the following:

$$p \vee q \tag{4.8}$$

is an example of a presupposition of question (4.3). Here are examples of presuppositions of question (4.4):

$$(p \wedge q) \vee (p \wedge \neg q) \tag{4.9}$$

$$p \wedge (q \vee \neg q) \tag{4.10}$$

$$q \vee \neg q \tag{4.11}$$

$$p \tag{4.12}$$

Let P be an arbitrary but fixed admissible partition. Observe that if a question is sound in P, then each presupposition of the question is true in P. On the other hand, the truth of a presupposition of a question need not warrant the soundness of the question. For instance, r is a presupposition of the following question of $\mathcal{L}^?_{cpl}$:

$$?\{p \wedge r, q \wedge r\} \tag{4.13}$$

but the question is not sound in an admissible partition in which r is true and both p and q are not true.

A presupposition whose truth warrants soundness of the question is called a prospective presupposition.

---
[5] See e.g. Belnap and Steel (1976), pp. 119-120. Belnap expresses this definition differently, however.

Definition 4.8 (Prospective presupposition). *A presupposition B of a question Q is prospective iff B $\mmodels$ dQ.*

A prospective presupposition is thus a presupposition that mc-entails the set of direct answers to the question. In other words, a prospective presupposition is a presupposition whose truth is both necessary and sufficient for the soundness of the question.

For example, the following d-wff:

$$r \wedge (p \vee q) \tag{4.14}$$

is a prospective presupposition of question (4.13), whereas the d-wff p is a prospective presupposition of the question:

$$?\{p \wedge q, p \wedge \neg q\} \tag{4.4}$$

Note that these are not the only prospective presuppositions of the analysed questions. For instance, (4.9) and (4.10) (but not (4.11)!) are also prospective presuppositions of question (4.4), while the set of prospective presupposition of question (4.13) includes, int. al., the d-wff:

$$(p \wedge r) \vee (q \wedge r) \tag{4.15}$$

In general, a prospective presupposition of a question of the form:

$$?\{A_1, \ldots, A_n\}$$

(of $\mathcal{L}^?_{cpl}$ or $\mathcal{L}^?_{fom}$) is either a d-wff of the form:

$$A_1 \vee \ldots \vee A_n \tag{4.16}$$

or a d-wff which is equivalent to it (by "equivalence" we mean here mutual entailment in a language). One can show that the existential generalization:

$$\exists x A x \tag{4.17}$$

is a prospective presupposition of the corresponding existential which-question:

$$?\mathbf{S}(Ax)$$

of $\mathcal{L}^?_{fom}$ (recall that admissible partitions of $\mathcal{L}^?_{fom}$ are determined by models, in which each individual has a name). The remaining prospective presuppositions of ?$\mathbf{S}(Ax)$ are equivalent to (4.17).

We do not claim, however, that each question of any language of the considered kind has prospective presupposition(s). On the contrary, there are languages which involve questions that lack prospective presuppositions. For example, when we have a general which-question of $\mathcal{L}^?_{fom}$ of the form:

$$?\mathbf{U}(Px)$$

where P represents an unary predicate, the existential generalization $\exists x Px$ is the strongest presupposition of the question, but is still not a prospective presupposition.

Notation. The set of presuppositions of a question Q will be referred to as $\mathsf{Pres}Q$, whereas the set of prospective presuppositions of Q will be designated by $\mathsf{PPres}Q$.

4.4 Normal questions and regular questions

The soundness of a question yields that each presupposition of the question, if there is any[6], is true. Yet, the converse need not hold. If it (non-vacuously) holds, the question is called normal. When there is a single presupposition whose truth warrants the soundness of a question, the question is called regular.

Normal question can be formally defined as follows.

Definition 4.9 (Normal question). *A question Q is normal iff $\mathsf{Pres}Q \neq \emptyset$ and $\mathsf{Pres}Q \models \mathsf{d}Q$.*

The condition $\mathsf{Pres}Q \neq \emptyset$ is dispensable if the intersection of the sets of "truths" of all the admissible partitions of a language is non-empty.

Regularity is a special case of normality.

Definition 4.10 (Regular question). *A question Q is regular iff there exists $B \in \mathsf{Pres}Q$ such that $B \models \mathsf{d}Q$.*

Contrary to appearance, these concepts are not equivalent. Since mc-entailment need not be compact (recall that we consider here an arbitrary language from a class of formal languages!), normality is not tantamount to regularity. The basic intuition which underlies the concept of regularity is: there exists a single presupposition of a question whose truth guarantees the existence of a true direct answer to the question. We have:

Corollary 4.11. *A question Q is regular iff $\mathsf{PPres}Q \neq \emptyset$.*

Remark. Regularity and normality are semantic concepts. One cannot say that a question (syntactically construed) is normal/regular in an "absolute" sense. For instance, existential which-questions are normal (and regular) in $\mathcal{L}^?_{fom}$, but would cease to be normal when admissible partitions were determined by all models of the language. If we enriched $\mathcal{L}^?_{fom}$ with a quantifier "there exist finitely many", general which-questions would become regular (provided that admissible partitions were defined as before).

4.5 Self-rhetorical questions and proper questions

The next concept needed is self-rhetoricity.

Definition 4.12 (Self-rhetorical question). *A question Q is self-rhetorical iff $\mathsf{Pres}Q \models A$ for some $A \in \mathsf{d}Q$.*

---

[6] Presuppositions are d-wffs. One can imagine a language of the considered kind in which there are no d-wffs entailed by each direct answer to a certain question and therefore its set of presuppositions is empty.

Observe that questions can be self-rhetorical for diverse reasons. If a question has a valid (i.e. true in each admissible partition of the language) d-wff among its direct answers, it is self-rhetorical. If all the direct answers to a question are equivalent and hence the "choice" offered by the question is only apparent, the question is self-rhetorical as well. A much weaker condition is also sufficient for self-rhetoricity: there is a direct answer which is entailed by all the other direct answers. If this is the case, the sets of direct answers and of presuppositions are not disjoint. Here is an example taken from the language $\mathcal{L}^?_{cpl}$:

$$?\{p, q, p \vee q\} \tag{4.18}$$

Again, self-rhetoricity is not an "absolute" property. For instance, the following:

$$?\{\Box p, \Box\Box p\} \tag{4.19}$$

is a self-rhetorical question of $\mathcal{L}^?_{S4}$, but ceases to be self-rhetorical when the underlying modal logic of d-wffs is weaker than K4.[7]

Finally, we introduce the concept of proper question.

Definition 4.13 (*Proper question*). *A question Q is proper iff Q is normal, but not self-rhetorical.*

Generally speaking, a question is proper just in case the truth of all its presuppositions warrants the existence of a true direct answer to it, but does not warrant the truth of any particular direct answer to the question.

4.6 Relative soundness

Let us now introduce:

Definition 4.14 (*Relative soundness*). *We say that a question Q is sound relative to a set of d-wffs X iff $X \models dQ$.*

Thus Q is sound relative to X if, and only if Q has a true direct answer in every admissible partition of the language in which all the d-wffs in X are true. In other words, it is impossible that all the d-wffs in X are true, but no direct answer to Q is true: if only X consists of truths, Q must be sound.

A warning is in order. One should not confuse relative soundness with soundness in a partition. Relative soundness is a *relation* between a question and a set of d-wffs. Soundness in a partition is a *property* which a question has or does not have.

One can easily prove:

Corollary 4.15. *A question Q is safe iff Q is sound relative to the empty set.*

Corollary 4.16. *A question Q is normal iff $\mathsf{Pres}Q \neq \emptyset$ and Q is sound relative to $\mathsf{Pres}Q$.*

---

[7] To be more precise, when $\Box p \rightarrow \Box\Box p$ is not a thesis of the logic.

4.7 Types of answers

So far we have operated with only one category of answers, that is, direct answers. However, the conceptual apparatus of MiES allows us to define further types of answers. We will restrict ourselves to categories which are most useful for our further considerations.

4.7.1 Just-complete answers

We start with just-complete answers. The following definition introduces a concept that is not superfluous when direct answers are defined syntactically.

Definition 4.17 (*Just-complete answer*). A d-wff C is a *just-complete answer to a question* Q iff $C \notin \mathsf{d}Q$, and for some $A \in \mathsf{d}Q$, both $C \models A$ and $A \models C$ hold.

Roughly, just-complete answers are equivalent to direct answers, but are not direct answers. By equivalence we mean mutual entailment.

It is convenient to introduce the following notational convention:

$$[\mathsf{d}Q] = \{C\colon \text{for some } A \in \mathsf{d}Q,\ C \models A \text{ and } A \models C\}$$

The set $[\mathsf{d}Q]$ comprises the direct answers to Q and the just-complete answers to Q.

4.7.2 Partial answers

Partial answers are defined by:

Definition 4.18 (*Partial answer*). A d-wff B is a *partial answer to a question* Q iff $B \notin [\mathsf{d}Q]$, but for some non-empty proper subset Y of $\mathsf{d}Q$:

1. $B \mathrel{\|\!\!\models} Y$, and
2. for each $A \in Y$: $A \models B$.

A partial answer is a d-wff that is neither a direct answer nor a just-complete answer, but which is true if, and only if a true direct answer belongs to some specified *proper subset* of the set of all the direct answers to the question. In other words, a partial answer to Q is a d-wff which narrows down (in the sense of Definition 3.17) the set of direct answers to Q. Recall that narrowing down is not tantamount to elimination.

Examples of partial answers will be given in section 4.7.3 below.

Note that a binary question, that is, a question which has only two direct answers, has no partial answer (understood in the sense of the above definition). However, binary questions, as well as other questions, still have answers which are neither direct answers nor just-complete answers.

4.7.3 Eliminative answers

Generally speaking, an eliminative answer, if true, eliminates at least one of the "possibilities" offered by a question. More formally:

Definition 4.19 (*Eliminative answer*). *A d-wff B is an eliminative answer to a question Q iff*

1. $B \notin [dQ]$, *and*
2. $B \in \mathsf{T_P}$ *for some admissible partition* P, *and*
3. *there exists $A \in dQ$ such that B eliminates A.*

The concept of elimination is understood here in the sense of Definition 3.15. When the classical negation[8] occurs in a language, an eliminative answer can also be defined as a consistent d-wff which entails the negation of at least one direct answer, but is not equivalent to any direct answer. However, there are languages of the considered kind in which negation does not occur or is understood non-classically, and for this reason we have formulated the definition in the current form.[9]

Eliminative answers vs. partial answers. There are eliminative answers which are not partial answers, and there are partial answers which are not eliminative. For consider the following question of $\mathcal{L}^?_{cpl}$:

$$?\{p, q, r\} \tag{4.20}$$

The d-wff:

$$p \vee q \tag{4.21}$$

is a partial answer to question (4.20), but is not an eliminative answer to the question. The d-wff:

$$\neg r \tag{4.22}$$

is an eliminative answer to question (4.20), but is not a partial answer to it. So one cannot *identify* partial answers with eliminative answers. These categories are not disjoint, however. For example, in the case of question (4.2) of $\mathcal{L}^?_{cpl}$:

$$? \pm |p, q|$$

each of the d-wffs p, $\neg p$, q, $\neg q$ is both a partial answer and an eliminative answer to the question.

---

[8] More generally, a negation for which a counterpart of the "law of contradiction" holds.

[9] Observe that if clause (2) had been dropped, each "inconsistent" (i.e. false in any admissible partition) d-wff would have been an eliminative answer to a question that has no inconsistent d-wff among its direct answers. Still, any contingent d-wff not equivalent to a direct answer is an eliminative answer to a question which has inconsistent d-wff(s) among its direct answers. We take this for granted.

4.7.4 Corrective answers

Roughly, a corrective answer is a consistent d-wff which eliminates (in the sense of Definition 3.15) the set of direct answers. We express this intuition by:

Definition 4.20 (*Corrective answer*). *A d-wff B is a corrective answer to a question Q iff*

1. $B \in \mathsf{T_P}$ *for some admissible partition* P, *and*
2. B *eliminates* $\mathsf{d}Q$.

Thus if a corrective answer is true in an admissible partition, no direct answer is true in the partition.

If the classical negation occurs in a language considered and $\mathsf{Pres}Q \neq \emptyset$, clause (2) of Definition 4.20 can be replaced with "B entails the negation of a presupposition of Q".

Clearly, each corrective answer is an eliminative answer. On the other hand, there are eliminative answers that are not corrective in the sense of Definition 4.20. As an illustration, let us consider the following question of $\mathcal{L}^?_{cpl}$:

$$?\{p \wedge q, p \wedge r\} \qquad (4.23)$$

The d-wffs $\neg q$ and $\neg r$ are eliminative answers to question (4.23), but are not corrective answers to the question. Here are examples of corrective answers to question (4.23):

$$\neg p \qquad (4.24)$$

$$\neg(q \vee r) \qquad (4.25)$$

One can easily show that the set of partial answers to a question and the set of corrective answers to the question are disjoint.

4.8 The applicability issue

Providing an adequate semantic analysis of NLQ's is a difficult task. It is not by accident that theories of questions are still diverse.

However, Minimal Erotetic Semantics is, in a sense, neutral here. If only NLQ's are represented/formalized by e-formulas, some assignment of sentential ppa's to e-formulas is given (regardless of how ppa's are conceptualized in detail – in our case as direct answers – and how the assignment is determined/defined: syntactically, semantically, or both), and the class of admissible partitions of the relevant formal language is determined, the "erotetic" concepts defined within MiES become applicable. To be more precise, they are directly applicable to e-formulas with well-defined/determined sets of ppa's, and indirectly – to NLQ's represented by them.

Moreover, it is irrelevant whether e-formulas/questions have been introduced into a formal language according to the "define within" method or the "enrich with" method. In the former case the "original" semantic concepts pertaining to other formulas apply to e-formulas as well. But nothing prevents us from using, in addition, a new collection of concepts, the MiES-concepts. They would constitute the "second collection" of semantic concepts pertaining directly to e-formulas and indirectly to NLQ's. The situation is analogous when e-formulas are defined syntactically, but their semantic analysis is also provided. If, however, questions of a formal language are defined only syntactically, then, assuming that some assignment of ppa's to questions/e-formulas is given, MiES shows how to deal with questions at the semantic level without providing a semantics for questions themselves.

Further readings. For the reasons of space, we did not present here all the erotetic concepts already defined within MiES. Similarly, we skipped most of the corollaries characterizing relations between concepts. More information on MiES can be found in Wiśniewski (1996) and Wiśniewski (2001). A model--theoretic variant of MiES is presented in Wiśniewski (1997a), and in Chapter 4 of the book Wiśniewski (1995). Recently Michal Peliš in Peliš (2011) proposed an account of MiES based on the notion of model as the basic one. His account includes definitions of some new concepts as well as simplifications of definitions of certain old concepts.

Part II

Inferences

5
Inferential Erotetic Logic: An Introduction

By and large, Inferential Erotetic Logic (IEL for short) is a logic that analyses inferences in which questions perform the role of conclusions, and proposes criteria of validity for these inferences. The idea of IEL originates from late 1980s, but IEL was developed in depth in the 1990s[1] as an alternative to the received view in the logic of questions, which situated the answerhood problem in the center of attention, and to the Interrogative Model of Inquiry, elaborated by Jaakko Hintikka.

5.1 Erotetic inferences

IEL starts with a trivial observation that before a question is asked or posed, a questioner must arrive at it. In many cases arriving at questions resembles coming to conclusions: there are premises involved and some inferential thought processes take place. If we admit that a conclusion need not be "conclusive", we can say that sometimes questions play the role of conclusions. But questions can also perform the role of premises: it often happens that an agent arrives at a question when looking for an answer to another question. Thus the concept of an *erotetic inference* is introduced. As a first approximation an erotetic inference may be defined as a thought process in which one arrives at a question on the basis of some previously accepted declarative sentence or sentences and/or a previously posed question. There are erotetic inferences of (at least) two kinds: the key difference between them lies in the type of premises involved. In the case of *erotetic inferences of the first kind* the set of premises consists of declarative sentence(s) only, and an agent passes from it to a question. For example:

> Andrew always comes in time, but now he is late.

> What has happened to him?

---

[1] Cf. Wiśniewski (1986), Wiśniewski (1989), Wiśniewski (1990a), Wiśniewski (1990b), Wiśniewski (1991), Wiśniewski (1994a), Wiśniewski (1995), Wiśniewski (1996), Wiśniewski (2001).

The premises of an *erotetic inference of the second kind* consist of a question and possibly some declarative sentence(s). For instance:

Where did Andrew leave for?
If Andrew took his famous umbrella, then he left for London; otherwise he left for Paris or Moscow.

Did Andrew take his famous umbrella?

Erotetic inferences in which no declarative premise occurs can be regarded as a special case of erotetic inferences of the second kind. Here is an example of an appropriate erotetic inference which does not rely on any declarative premise:

Is Andrew silly and ugly?

Is Andrew ugly?

An inference, even erotetic, is always someone's inference. In its general setting, however, IEL abstracts from this: erotetic inferences are construed syntactically. Erotetic inferences of the first kind are viewed as ordered pairs $\langle X, Q \rangle$, where X is a finite and non-empty set of declarative sentences and Q is a question. Similarly, an erotetic inference of the second kind is identified with an ordered triple $\langle Q, X, Q_1 \rangle$, where Q, Q_1 are questions and X is a finite (possibly empty) set of declarative sentences. When formal languages with questions are dealt with, X is a set of declarative well-formed formulas. Erotetic inferences construed syntactically are also called *erotetic arguments*.

5.2 Validity of erotetic inferences

IEL proposes some conditions of validity of erotetic inferences.

As long as we are concerned with inferences which have only declaratives as premises and conclusions, validity amounts to the transmission of truth: if the premises are all true, the conclusion must be true as well. However, it is doubtful whether it makes any sense to assign truth or falsity to questions and thus one cannot apply the above concept of validity to erotetic inferences. But in the case of questions the concept of soundness seems to play an equally important role as the concept of truth in the realm of declaratives. Recall that a question Q is *sound* if at least one direct answer to Q is true, and unsound otherwise.

There are erotetic inferences of (at least) two kinds, and the conditions of validity are distinct for each kind.

5.2.1 Validity of erotetic inferences of the first kind

An erotetic inference of the first kind is conceived as an ordered pair $\langle X, Q \rangle$, where X is a finite and non-empty set of declarative sentences/d-wffs, and Q is a question. The elements of X are the premises, and Q is the conclusion.

5.2 Validity of erotetic inferences

It seems natural to impose the following necessary condition of validity on erotetic inferences of the first kind:

(C_1) (TRANSMISSION OF TRUTH INTO SOUNDNESS). *If the premises are all true, then the question which is the conclusion must be sound.*

Is this sufficient? Certainly not. For if (C_1) were sufficient, the following inferences would be valid:

Andrew is rich.
Andrew is happy.
___
Is Andrew happy?

If Andrew is rich, then he is happy.
Andrew is rich.
___
Is Andrew happy?

What is wrong with the above inferences? The question which is the conclusion has a direct answer which provides us with information that is already present (directly or indirectly) in the premises. In other words, the question which is the conclusion is logically redundant and thus not informative. So IEL imposes the following additional necessary condition of validity on erotetic inferences of the first kind:

(C_2) (INFORMATIVENESS). *A question which is the conclusion must be informative relative to the premises.*

Informativeness is then explicated as the lack of entailment of any direct answer from the premises; the applied concept of entailment need not be classical (see below).

Here are examples of valid erotetic inferences of the first kind:

Mary is Peter's mother.
If Mary is Peter's mother, then Bill is
Peter's father or George is Peter's father.
___
Who is Peter's father: Bill or George?

Someone stole the necklace.
___
Who did it?

5.2.2 Validity of erotetic inferences of the second kind

An erotetic inference of the second kind is conceived as an ordered triple $\langle Q, X, Q_1 \rangle$, where Q, Q_1 are questions and X is a finite (possibly empty) set of declarative sentences/d-wffs. The question Q is the *interrogative premise* (we will also call it *initial question*), the elements of X are *declarative premises*, and the question Q_1 is the conclusion.

The natural generalization of the standard condition of validity is:

($\mathbf{C}_3$) (TRANSMISSION OF SOUNDNESS/TRUTH INTO SOUNDNESS). *If the initial question is sound and all the declarative premises are true, then the question which is the conclusion must be sound.*

As an illustration, let us consider the following inference:

Where did Andrew leave for: Paris, London or Moscow?
If Andrew left for Paris, London or Moscow, then he departed in the morning or in the evening.
If Andrew departed in the morning, then he left for Paris or London.
If Andrew departed in the evening, then he left for Moscow.

When did Andrew depart: in the morning, or in the evening?

The question which is the interrogative premise, that is:

Where did Andrew leave for: Paris, London or Moscow? (5.1)

offers three "possibilities": Paris, London, and Moscow. The use of "leaves for" suggests that these possibilities are mutually exclusive. We can construe question (5.1) as having three direct answers, namely "Andrew left for Paris", "Andrew left for London", and "Andrew left for Moscow".[2] The disjunction of all of them is not a logical truth, however: it can happen that none of them is true. Andrew might have left for Rome, or stayed at home, and so forth. In other words, question (5.1) need not be sound.

Similarly, the question which is the conclusion, i.e.:

When did Andrew depart: in the morning, or in the evening? (5.2)

need not be sound either. The direct answers to question (5.2) are: "Andrew departed in the morning" and "Andrew departed in the evening". It may happen that neither of them is true.

Finally, the declarative premises of the above inference need not be true.

However, *suppose* that question (5.1) being the interrogative premise is sound, and that all the declarative premises are true. It follows that question (5.2) which is the conclusion must be sound: it is impossible, given the assumptions, that the direct answers are both false. In other words, the claim of condition ($\mathbf{C}_3$) is fulfilled in the analysed case.

Sometimes the claim is fulfilled for trivial reasons, as in:

---

[2] The short answers "Paris", "London" and "Moscow" have, respectively, the same meanings (in the current context) as the direct answers.

5.2 Validity of erotetic inferences

Where did Andrew leave for?
If Andrew took his famous umbrella, then he left for London;
otherwise he left for Paris or Moscow.

Did Andrew take his famous umbrella?

The question which is the conclusion must be sound because the set of direct answers consists of a sentence and its classical negation.

Let us stress that IEL regards (C_3) only as a necessary condition of validity of erotetic inferences of the second kind. Why? If condition (C_3) had been sufficient, then, for instance, the following would have been valid inferences:

Is Andrew a logician?
Some philosophers are logicians, and some are not.

Is Andrew a philosopher?

Is Coco a human?
Humans are mammals.

Is Coco a mammal?

The problem here is that the questions which are conclusions have direct answers that are cognitively useless: these answers, if accepted, would not contribute to finding answers to initial questions.[3] On the other hand, an intuitive account of validity suggests that direct answers to the question which is the conclusion should be potentially useful, on the basis of the declarative premises, for finding an answer to the initial question. To secure this, IEL imposes the following necessary condition of validity of erotetic inferences of the second kind:

(C_4) (OPEN-MINDED COGNITIVE USEFULNESS). *For each direct answer B to the question which is the conclusion there exists a non-empty proper subset Y of the set of direct answers to the initial question such that the following condition holds:*
  (♣) *if B is true and all the declarative premises are true, then at least one direct answer $A \in Y$ to the initial question must be true.*

In other words, each direct answer to the question which is the conclusion should, together with the declarative premises, narrow down the class of possibilities offered by the initial question. Condition (C_4) can be clarified in terms of multiple-conclusion entailment (see below). As a special case, a singleton class of possibilities can show up, but this is not required in general.

The following inference is valid:

---

[3] In the first case none of the answers is potentially useful. As for the second case, the negative answer is useful, whereas the affirmative answer is useless. Needless to say, in any of the above cases condition (C_3) is satisfied for a trivial reason, due to the structure of the "question-conclusion" only.

How old is Andrew?
Andrew is as old as Peter.
―――――――――――――――――
How old is Peter?

Condition (C_4) is now fulfilled in the "one-to-one" way: each instantiation of "Peter is ... years old", together with the declarative premise, yield a (direct) answer to the initial question. In other words, the relevant proper subsets are singleton sets. It is obvious that condition (C_3) is fulfilled as well.

Now consider the already analysed inference:

Where did Andrew leave for: Paris, London or Moscow?
If Andrew left for Paris, London or Moscow, then he departed in the morning or in the evening.
If Andrew departed in the morning, then he left for Paris or London.
If Andrew departed in the evening, then he left for Moscow.
―――――――――――――――――
When did Andrew depart: in the morning, or in the evening?

The direct answers to the question which is the conclusion are:

$$\text{\textit{Andrew departed in the morning.}} \tag{5.3}$$

$$\text{\textit{Andrew departed in the evening.}} \tag{5.4}$$

Suppose that answer (5.3) is true and that all the declarative premises are true. It follows that Andrew left for Paris or London. In other words, given the assumptions, a true direct answer to the initial question belongs to the following proper subset of the set of all direct answers:

$$\{\textit{Andrew left for Paris, Andrew left for London}\} \tag{5.5}$$

Now suppose that answer (5.4) is true and that the declarative premises are true. It follows that Andrew left for Moscow. In terms of sets: a true direct answer to the question which is the conclusion belongs to the following proper subset of the set of all the direct answers:

$$\{\textit{Andrew left for Moscow}\} \tag{5.6}$$

which happens to be a singleton set. Thus condition (C_4) is fulfilled. We have already shown that condition (C_3) is satisfied as well. Hence the analysed inference is valid.

The already analysed inference:

Where did Andrew leave for?
If Andrew took his famous umbrella, then he left for London; otherwise he left for Paris or Moscow.
―――――――――――――――――
Did Andrew take his famous umbrella?

is valid as well. Condition ($\mathbf{C_3}$) is fulfilled for trivial reasons. As for the condition ($\mathbf{C_4}$), the affirmative answer to the question which is the conclusion yields, together with the declarative premise, that Andrew left for London. Thus the proper subset (of the set of direct answers to the initial question) that corresponds to the affirmative answer is:

$$\{Andrew\ left\ for\ London\} \tag{5.7}$$

In the case of the negative answer, the corresponding proper subset is:

$$\{Andrew\ left\ for\ Paris,\ Andrew\ left\ for\ Moscow\} \tag{5.8}$$

Finally, the following is also a valid inference:

Where does Andrew live?
Andrew lives in a university town in Western Poland.

Which towns in Western Poland are university towns?

The conclusion is an open-condition question understood in the "all-the-cases" way.[4] The question should be formalized as a general which-question[5]; a direct answer to it specifies a finite list of towns and claims, truly or not, that these are the only university towns in Western Poland. So each direct answer to the question which is the conclusion narrows down, along with the declarative premise, the class of possibilities offered by the initial question and thus condition ($\mathbf{C_4}$) is fulfilled. It is clear that condition ($\mathbf{C_3}$) is satisfied as well.

Comments. Questions usually have many direct answers. Let us stress that condition ($\mathbf{C_4}$) is rather demanding: it is required that for *each* direct answer to the "question-conclusion" there exists a non-empty proper subset of the set of direct answers to the initial question such that the relevant clause ($\clubsuit$) is satisfied. Roughly, each direct answer to a "question-conclusion" of a valid erotetic inference (of the second kind) is supposed to be potentially cognitively useful. One may argue that this is too much and that only some of the answers should do. However, when each direct answer to a "question-conclusion" is potentially cognitively useful, then a true answer, if found, becomes actually cognitively useful: the "search space" for a true direct answer to the initial question decreases.[6] Condition ($\mathbf{C_4}$) is neutral with respect to the issue which (if any) direct answer to the "question-conclusion" is actually true: it only requires each direct answer to the question to stay in the appropriate semantic relation with some non-empty proper subset of the set of all the direct answers to the initial question. This is why the label "open-minded" is used in the name of condition ($\mathbf{C_4}$).

---

[4] See Chapter 1, section 1.1.
[5] See Chapter 2, section 2.4.3.
[6] Of course, assuming that the declarative premises are true. The initial "search space" is the whole set of direct answers to the initial question.

5.3 Validity and question raising

Validity is a normative notion. In the case of erotetic inferences the appropriate notion of validity is given neither by God nor by Tradition. So some more or less arbitrary decisions have to be made. IEL decides to consider as valid these erotetic inferences of the first kind which have the features described by conditions (C_1) and (C_2). As for erotetic inferences of the second kind, conditions (C_3) and (C_4), jointly, determine the relevant concept of validity used within IEL.

Conditions (C_1) – (C_4) are expressed somewhat loosely. Of course, IEL offers more than just formulating them. The following semantic concepts are introduced: (i) *evocation* of questions by sets of d-wffs, and (ii) *erotetic implication* of questions by questions and (possibly) sets of d-wffs. Validity of erotetic inferences of the first kind is then defined in terms of evocation, whereas validity of erotetic inferences of the second kind is defined by means of erotetic implication. The proposed definitions of evocation and erotetic implication are explications of the relevant notions of question raising (cf. Wiśniewski (1995), Chapter 1). By defining the semantic concept "a set of d-wffs X evokes a question Q" we explicate the concept "a question Q arises from a set of declarative sentences X". The definition of "a question Q implies a question Q_1 on the basis of a set of d-wffs X" provides an explication of the notion "a question Q_1 arises from a question Q and a set of declarative sentences X". Thus, although conditions (C_1) and (C_2) on the one hand, and conditions (C_3) and (C_4) on the other seem diverse at first sight, the analysis of validity of erotetic inferences proposed by IEL is based on a certain general idea: *the conclusion of a valid erotetic inference arises from the premises.*

5.4 The logical basis of IEL

5.4.1 Syntax

IEL accepts the non-reductionist approach to NLQ's. As for formal languages, questions are supposed to be introduced into them according to the "enrich with" method (see section 2.1 of Chapter 2). What we need is a formal language which has both d-wffs and e-formulas/questions among its well-formed expressions, where questions are distinct from well-formed expressions of other categories. Then we need some assignment of direct answers to questions. It is assumed that questions (but not necessarily all of them) of formal languages represent questions of natural languages: *question Q represents a NLQ Q^* construed in such a way that possible just-sufficient answers to Q^* are represented/formalized by direct answers to Q*, where "just-sufficient" means "satisfying the request of a question by providing neither less nor more information than it is requested". The exemplary languages presented in Chapter 2, built according to the semi-reductionistic pattern, are of the above kind, but they are not the only one. One can construct appropriate languages e.g. according to Kubiński's approach or Belnap's approach, to mention only the richest sources of ideas (see Belnap and Steel (1976), Kubiński (1980)).

Some specific assumptions. In general considerations concerning IEL it is assumed that the following conditions are satisfied by the relevant formal languages with questions:

(sc$_1$) *direct answers are sentences, i.e. d-wffs with no individual or higher-order free variables (since sentential functions are not definite enough in order to answer anything);*[7]

(sc$_2$) *each question has at least two direct answers (since a necessary condition of being a question is to present at least two "alternatives" or possibilities).*

Sometimes the following condition is also imposed:

(sc$_3$) *each finite and at least two-element set of sentences is the set of direct answers to some question.*

Note that all the languages considered in sections 2.4.1, 2.4.2 and 2.4.3 of Chapter 2 fulfil conditions (sc$_1$) and (sc$_2$). Condition (sc$_3$) is not fulfilled by the erotetic sequent language $\mathcal{L}^?_{\vdash cpl}$, but is satisfied by the remaining languages.

5.4.2 Semantics

Both evocation and erotetic implication are semantic concepts, defined inter alia in terms of mc-entailment and entailment. Thus it is supposed that the declarative part of the language under consideration is supplemented with a semantics rich enough to define these concepts of entailment. In the general setting we use the conceptual apparatus of Minimal Erotetic Semantics described in Chapters 3 and 4; we suppose that the class of admissible partitions of a language is determined and this, as we have seen, enables us to define – and operate with – the remaining concepts.

Needless to say, the conceptual apparatus of MiES is general enough to allow both Classical Logic and a non-classical logic to be the logic of d-wffs. IEL is neutral with respect to the issue of what "The Logic" of declaratives is.

Final remark. There are erotetic inferences (of the first and second kind) which are beyond the scope of the analysis of validity provided by IEL. The conditions of validity ($\mathbf{C}_1$) – ($\mathbf{C}_4$) presuppose that the relevant questions have well-defined sets of direct answers, and that each direct answer has a truth value. So one cannot apply them to erotetic inferences whose conclusions (and erotetic premises) do not satisfy these assumptions. The problem of validity of erotetic inferences involving such questions is left open by IEL.

---

[7] As for propositional languages, direct answers are propositional formulas.

6

Evocation of Questions

In this chapter we introduce the concept of evocation of a question by a set of d-wffs, and we address the issue of validity of erotetic inferences of the first kind.

On the whole we assume that we deal with a formal language with questions, $\mathcal{L}$, which satisfies the conditions specified in section 5.4 of Chapter 5. The semantic concepts used are characterized in Chapters 3 and 4, and are understood accordingly. For brevity, we omit the specifications "in $\mathcal{L}$" and "of $\mathcal{L}$".

6.1 Definition of evocation

The basic intuition which underlies the concept of evocation is very simple. Let X be a set of d-wffs. If the truth of all the d-wffs in X guarantees the existence of a true direct answer to a question Q, but does not yield the truth of any single direct answer to Q, we say that X evokes Q. Of course, we do not restrict ourselves to cases when X consists of truths: we only claim that Q is sound (i.e. has a true direct answer) *assuming that* X consists of truths, and that this assumption is not "strong enough" to determine which answer to Q is true: only the existence of a true (direct) answer is guaranteed.

The concept of multiple-conclusion entailment (see section 3.2 of Chapter 3) seems ideally suited to express the above idea in exact terms. Let us recall that it may happen that a set of d-wffs X mc-entails a set of d-wffs Y, although no single d-wff in Y is entailed by X. Suppose that X and Y are connected that way. Thus, intuitively speaking, we have the following situation: the truth of all the d-wffs in X guarantees the existence of a true d-wff in Y, but does not guarantee the truth of any particular element of Y. In other words: the truth of all the X's yields that at least one element of Y is true, but does not determine which one is true. Now think of Y as of the set dQ of direct answers to a question Q, and the definition of evocation is ready. To be more precise, we put:

6 Evocation of Questions

Definition 6.1 (*Evocation of questions*). *A set of d-wffs X evokes a question Q (in symbols: $\mathbf{E}(X,Q)$) iff*

1. $X \mathrel{||\!\!=} \mathsf{d}Q$, *and*
2. *for each $A \in \mathsf{d}Q : X \mathrel{||\!\!\not=} \{A\}$.*

Clause (2) is formulated in terms of mc-entailment for uniformity only: according to Corollary 3.14 (see page 34), X does not mc-entail $\{A\}$ if, and only if X does not entail A.

Strictly speaking, Definition 6.1 provides us with a *schema* of definition of evocation. Recall that we have assumed that an arbitrary but fixed formal language with questions that fulfils the conditions described in section 5.4 of Chapter 5 is considered. When a language is specified, we are confronted with *evocation in the language*. To highlight this, one can add a subscript to $\mathbf{E}$. Thus, for example, $\mathbf{E}_{\mathcal{L}^?_{cpl}}$ and $\mathbf{E}_{\mathcal{L}^?_{fom}}$ refer to evocation in the languages $\mathcal{L}^?_{cpl}$ and $\mathcal{L}^?_{fom}$, respectively.

6.1.1 Transmission of truth into soundness

Clause (1) of Definition 6.1 expresses in exact terms the idea of transmission of truth into soundness: if only all the d-wffs in X are true, then Q must be sound. "Must" means here: there is no admissible partition $\mathsf{P} = \langle \mathsf{T}_\mathsf{P}, \mathsf{U}_\mathsf{P} \rangle$ such that $X \subset \mathsf{T}_\mathsf{P}$ and $\mathsf{d}Q \cap \mathsf{T}_\mathsf{P} = \emptyset$. In other words, an evoked question is supposed to be sound relative to the evoking set. Of course, we do not require that, in each admissible partition, X consists of truths. We only require that the condition:

$$\text{if } X \subset \mathsf{T}_\mathsf{P}, \text{ then } \mathsf{d}Q \cap \mathsf{T}_\mathsf{P} \neq \emptyset$$

holds for every admissible partition P.

Assume that $\mathbf{E}(X,Q)$. Tables (6.1) and (6.2) describe possible semantic connections between X and Q. $\mathsf{P} = \langle \mathsf{T}_\mathsf{P}, \mathsf{U}_\mathsf{P} \rangle$ stands for an arbitrary but fixed admissible partition of $\mathcal{L}$; a question is called *unsound* in P if the question is not sound in P.

Table 6.1. *From evoking d-wffs to evoked question.*

X	Q
$X \subset \mathsf{T}_\mathsf{P}$	sound in P
$X \not\subset \mathsf{T}_\mathsf{P}$	sound in P or unsound in P

Table 6.2. *From evoked question to evoking d-wffs.*

Q	X
sound in P	$X \subset \mathsf{T}_\mathsf{P}$ or $X \not\subset \mathsf{T}_\mathsf{P}$
unsound in P	$X \not\subset \mathsf{T}_\mathsf{P}$

6.1.2 Informativeness

Clause (2) of Definition 6.1 explicates the idea of informativeness of a question being the conclusion.[1] According to the clause, informativeness amounts to the lack of entailment of any direct answer to Q from X. The underlying idea is: the problem expressed by Q cannot be resolved just by performing a legitimate deduction of a direct answer to Q from the set X. Or, what is the other side of the same coin, one can legitimately derive *more* from the set X enriched with a direct answer to Q than from the set X itself, for any direct answer to Q. But, as "is entailed" is neither synonymous with "is known as entailed" nor with "is entailed and known", the degree of idealization (with respect to the phenomenon of question raising) is high.[2]

6.2 Some properties of evocation

For properties of evocation see Wiśniewski (1991), Wiśniewski (1995), Wiśniewski (1996). Let us only mention some of them.

The corollaries presented below are immediate consequences of the relevant definitions.

Corollary 6.2. *Let Q be a normal question. Then $\mathbf{E}(X, Q)$ iff $X \models B$ for each $B \in \mathsf{Pres}Q$, and $X \not\models A$ for each $A \in \mathsf{d}Q$.*

Therefore evocation of normal questions amounts to the fact that each presupposition is entailed, but no direct answer is entailed. The situation is even simpler in the case of regular questions, that is, questions which have single presuppositions whose truth guarantee their soundness, i.e. prospective presuppositions. The following holds:

Corollary 6.3. *Let Q be a regular question. Then $\mathbf{E}(X, Q)$ iff $X \models B$ for some $B \in \mathsf{PPres}Q$, and $X \not\models A$ for each $A \in \mathsf{d}Q$.*

A question is proper if it is normal but not self-rhetorical.[3] We have:

Corollary 6.4. *A question Q is proper iff $\mathsf{Pres}Q \neq \emptyset$ and $\mathbf{E}(\mathsf{Pres}Q, Q)$.*

Evocation is not monotone. The following hold:

---

[1] See section 5.2.1 of Chapter 5.
[2] Another problem with the definition of evocation is this. Suppose that the underlying entailment relation is only "positively decidable", that is, roughly, there is an effective method of deciding that entailment holds, but there is no effective method of deciding that it does not hold (to be more precise: entailment is r.e., but is not recursive). In such a case evocation is not even positively decidable. However, we can still prove that some questions of a given form are evoked by sets of d-wffs that comprise d-wffs of strictly defined forms, and results of this kind enable us to formulate question-evoking rules. These rules can be claimed as governing valid erotetic inferences.
[3] See section 4.5 of Chapter 3.

6 Evocation of Questions

Corollary 6.5. *Let* $\mathbf{E}(X,Q)$ *and* $X \subset Y$. *Then* $\mathbf{E}(Y,Q)$ *iff* $Y \not\models A$ *for each* $A \in \mathsf{d}Q$.

Corollary 6.6. *Let* $\mathbf{E}(X,Q)$ *and* $\mathsf{d}Q \subset \mathsf{d}Q_1$. *Then* $\mathbf{E}(X,Q_1)$ *iff* $X \not\models A$ *for each* $A \in \mathsf{d}Q_1 \setminus \mathsf{d}Q$.

Finally, let us note:

Corollary 6.7. *If* $\mathbf{E}(X,Q)$, *then* $X \subset \mathsf{T}_\mathsf{P}$ *for some admissible partition* P.

Contrary to appearance, Corollary 6.7 is not tantamount to the claim that inconsistent sets of d-wffs do not evoke any question. The reason is this: nothing prevents us from taking a paraconsistent logic as a background. In such a case (at least some) classically inconsistent sets will belong to sets of "truths" of appropriate admissible partitions and *Ex Falso Quodlibet* will not hold. The general framework of IEL does not prejudge what admissible partitions are and thus the underlying logic of declaratives need not be classical.

6.2.1 A digression: Meheus' analysis

The case of inconsistent premises is a challenge to the analysis of question raising, since questions often arise from inconsistencies. The solution sketched above is of course not the only one. As it is pointed at in Meheus (1999):

> *When generalizing the idea of erotetic arguments to the inconsistent case, a distinction has to be made between two types of situations: those in which the inconsistencies are (for the time being) accepted as true, and those in which this is not the case. (...) Important for this [second] type is that the requirements for erotetic arguments have to be defined not with respect to the inconsistent set* Γ, *but with respect to the consistent core of* Γ. (Meheus (1999), p. 58)

Meheus (1999) analyses the second situation and proposes interesting solutions. The starting point is this: instead of considering a (possibly inconsistent) set of premises Γ, one considers a couple of sets of d-wffs, $\{\Gamma_0, \Gamma_1\}$. Intuitively, Γ_0 represents premises which, due to some external reasons, are accepted as true, whereas Γ_1 is made up of premises which, again for external reasons, are interpreted as only possibly true. For brevity, let us consider only the propositional case, and only one of the logics proposed by Meheus.

The analysis goes on in terms of the modal propositional logic S5. One considers S5-models of the form:

$$\langle W, \{w_0\}, V \rangle \tag{6.1}$$

where W is a non-empty set (of possible worlds), $w_0 \in W$, and V is the valuation function defined in the usual way; in this section by a S5-model we will mean a structure of the above kind.[4] A model (6.1) *verifies* a d-wff A iff

---

[4] Since we deal with S5, we do not need the accessibility relation. Let us stress that we depart from the S5-semantics used in the original paper. Moreover, Meheus works in the framework of adaptive logics; however, for the reasons of space, we do not present her proposal in this setting.

$V(A, w_0) = 1$. Let $\Delta = \{\Gamma_0, \Gamma_1\}$, where Γ_0 and Γ_1 are sets of non-modal d-wffs. We say that (6.1) is a model of $\Delta = \{\Gamma_0, \Gamma_1\}$ iff for each $A \in \Gamma_0$ we have $V(A, w_0) = 1$, and for each $B \in \Gamma_1$ we have $V(\Diamond B, w_0) = 1$. A d-wff A is S5-*entailed* by $\Delta = \{\Gamma_0, \Gamma_1\}$ if A is verified by each model of Δ. A *Dab*-formula is a disjunction of d-wffs of the form $\Diamond C \wedge \neg C$, where C is a literal, that is, a propositional variable or a negation of a propositional variable. Generally speaking, a *Dab*-formula expresses the fact that the relevant literal(s) "behave(s) abnormally". Now we are ready to define the crucial concept.

A model $\boldsymbol{M} = \langle W, \{w_0\}, V \rangle$ of Δ is *reliable* iff the following condition holds:
($\star$) for each literal C: if $\Diamond C \wedge \neg C$ is verified by $\boldsymbol{M}$, then $\Diamond C \wedge \neg C$ is a disjunct of a certain minimal *Dab*-formula which is S5-entailed by Δ.

(The "minimality" condition means that no formula which results from the *Dab*-formula by dropping a disjunct is S5-entailed by Δ.) Then mc-entailment between couples of sets of non-modal d-wffs and sets of non-modal d-wffs is defined in terms of reliable models, and evocation of questions is defined according to the schema presented in Definition 6.1. The only difference is that questions are now evoked, strictly speaking, by couples of sets of d-wffs $\{\Gamma_0, \Gamma_1\}$.

The outcome of Meheus' analysis is this. When $\Gamma_0 \cup \Gamma_1$ is a (classically) consistent set, we get all the properties (and examples) of evocation defined in the standard setting. If, however, $\Gamma_0 \cup \Gamma_1$ is inconsistent, it still evokes some questions. The same holds for the remaining two proposals (which differ from the just presented in defining the relevant class of models), included in Meheus (1999).

6.2.2 Generation of questions

Generation of questions is a special case of evocation: we say that a set of d-wffs X generates a question Q iff X evokes Q and the set of direct answers to Q is not mc-entailed by the empty set. The underlying intuition is: a generated question is *made sound* by the generating set and is informative with respect to this set. For generation of questions see Wiśniewski (1989), Wiśniewski (1990b), Wiśniewski (1991), Wiśniewski (1995).

6.3 Examples of evocation

Let us now present some examples of evocation.

Recall that $\mathcal{L}^?_{cpl}$ is the language of CPL enriched with questions (see section 2.4.1 of Chapter 2 for details). The letters p, q, r stand for propositional variables. For brevity, we use object-level language expressions instead of their metalinguistic names, and we simply list the elements of evoking sets. For conciseness, we write $\mathbf{E}$ instead of $\mathbf{E}_{\mathcal{L}^?_{cpl}}$. Here are examples of evocation in $\mathcal{L}^?_{cpl}$.

$$\mathbf{E}(p \vee \neg p, ?p) \qquad (6.2)$$

$$\mathbf{E}(p \vee q, ?p) \qquad (6.3)$$

64   6 Evocation of Questions

$$\mathbf{E}(p \vee q, ?q) \tag{6.4}$$
$$\mathbf{E}(p \to q, ?p) \tag{6.5}$$
$$\mathbf{E}(p \to q, ?q) \tag{6.6}$$
$$\mathbf{E}(p \leftrightarrow q, ?\{p \wedge q, \neg p \wedge \neg q\}) \tag{6.7}$$
$$\mathbf{E}(p \vee q, ?\{p, q\}) \tag{6.8}$$
$$\mathbf{E}(p \vee q, ?(p \wedge q)) \tag{6.9}$$
$$\mathbf{E}(p \vee q, ?\{p \wedge q, p \wedge \neg q, \neg p \wedge q\}) \tag{6.10}$$
$$\mathbf{E}(p \to q \vee r, ?\{p \to q, p \to r\}) \tag{6.11}$$
$$\mathbf{E}(p \to q \vee r, p, ?\{q, r\}) \tag{6.12}$$
$$\mathbf{E}(\neg(q \wedge r), ?\{\neg q, \neg r\}) \tag{6.13}$$
$$\mathbf{E}(p \wedge q \to r, ?\{p \to r, q \to r\}) \tag{6.14}$$
$$\mathbf{E}(p \wedge q \to r, \neg r, ?\{\neg p, \neg q\}) \tag{6.15}$$
$$\mathbf{E}((p \vee q) \vee r, ?\{p, q \vee r\}) \tag{6.16}$$
$$\mathbf{E}(p \wedge (q \vee r), ?\{p \wedge q, p \wedge r\}) \tag{6.17}$$
$$\mathbf{E}(p \wedge (q \vee r), ?\{(p \wedge q) \wedge \neg r, (p \wedge r) \wedge \neg q, p \wedge (q \wedge r)\}) \tag{6.18}$$

Now we turn to the language $\mathcal{L}^?_{fom}$ described in section 2.4.3 of Chapter 2; the semantics is presented in section 3.1.6 of Chapter 3. Recall that admissible partitions of $\mathcal{L}^?_{fom}$ are determined by those models of $\mathcal{L}^?_{fom}$ in which each element of the domain has a name. The letters P, R are used as metalanguage variables for predicates, and c, c^*, c_1, ... are metalanguage variables for individual constants. Distinct metalanguage variables are supposed to represent distinct object-level language entities.

Here are examples of evocation in $\mathcal{L}^?_{fom}$. For brevity, we use $\mathbf{E}$ instead of $\mathbf{E}_{\mathcal{L}^?_{fom}}$.

$$\mathbf{E}(Pc_1, \ldots, Pc_n, ?\forall x Px) \tag{6.19}$$
$$\mathbf{E}(Pc_1 \wedge Rc_1, \ldots, Pc_n \wedge Rc_n, ?\forall x(Px \to Rx)) \tag{6.20}$$
$$\mathbf{E}(Pc_1 \wedge Rc_1, \ldots, Pc_n \wedge Rc_n, Pc_{n+1}, ?Rc_{n+1}) \tag{6.21}$$
$$\mathbf{E}(Pc_1, Pc_2, ?c_1 = c_2) \tag{6.22}$$
$$\mathbf{E}(\exists x Px, ?\{Pc, \exists x(Px \wedge x \neq c)\}) \tag{6.23}$$
$$\mathbf{E}(\exists x(Px \wedge (x = c_1 \vee \ldots \vee c_n)), ?\{Pc_1, \ldots, Pc_n\}), \tag{6.24}$$

where $n > 1$.

$$\mathbf{E}(\neg \forall x(x = c_1 \vee \ldots \vee x = c_n \to Px), ?\{\neg Pc_1, \ldots \neg Pc_n\}), \tag{6.25}$$

where $n > 1$.

$$\mathbf{E}(\exists x(Px \wedge (x = c_1 \vee \ldots \vee x = c_n)), ?\mathbf{S}(Px)), \tag{6.26}$$

where $n > 1$.

$$\mathbf{E}(\neg \forall x(x=c_1 \vee \ldots \vee x=c_n \rightarrow Px), ?\mathbf{S}(\neg Px)), \qquad (6.27)$$

where $n > 1$.

$$\mathbf{E}(\exists x Px, ?\mathbf{S}(Px)) \qquad (6.28)$$

$$\mathbf{E}(\forall x(Px \leftrightarrow x=c_1 \vee x=c_n) \vee \forall x(Px \leftrightarrow x=c_1^* \vee \ldots \vee x=c_k^*), ?\mathbf{U}(Px)) \quad (6.29)$$

It is not by accident that a general which-question occurs above only once. In order to analyse evocation of these questions one needs, besides mc-entailment relativized to models in which all elements of domains have names in the language, also the quantifier "there exist finitely many" (cf. Wiśniewski (1990a)), or at least numerical quantifiers (cf. Wiśniewski (1995)).

Remark. We have considered above only relatively simple languages enriched with rather simple questions. For further examples of evocation, in particular evocation of complex questions in more sophisticated first-order languages enriched with questions see Wiśniewski (1995), Chapter 5. For evocation of questions based on non-classical logics see, besides Meheus (1999), also Meheus (2001), De Clercq and Verhoeven (2004), De Clercq (2005).

6.4 Evocation and validity

We are now ready to define validity of erotetic inferences of the first kind, that is, inferences leading from declarative sentences/d-wffs to questions. The clauses of the definition of evocation explicate the intuitive conditions of validity ($\mathbf{C}_1$) and ($\mathbf{C}_2$) specified in section 5.2.1 of Chapter 5. Thus we put:

Definition 6.8 (*Validity I*). *An erotetic inference of the first kind, $\langle X, Q \rangle$, is valid iff $\mathbf{E}(X, Q)$.*

By proposing the above definition we implicitly assume that the X's and Q are expressions of a language for which evocation is appropriately defined.

The second clause of the definition of evocation amounts to the lack of entailment of directs answers to the evoked question from the evoking set. On the other hand, the intuitive concept of informativeness of a question with regard to the relevant premises can be construed in a less demanding way: one can require that no direct answer is an "immediate" or "obvious" consequence of the premises. It is still an open problem how to express the idea of "not being an immediate consequence" in purely *semantic* terms. Whatever the solution might have been, applying it in the definition of validity would probably resulted in a complex notion lacking intuitive clarity. This is the reason for which IEL defines validity of erotetic inferences of the first kind in terms of question evocation, where informativeness is conceptualized as the lack of entailment.

6.4.1 Some comments

Definition 6.8 pertains to erotetic inferences in formal languages, since, strictly speaking, evocation is defined only for these languages. On the other hand, erotetic inferences are most often performed by means of premises and conclusions being expressions of natural languages. The above concept of validity can be applied to such inferences only in an indirect way: an inference is valid just in case evocation holds between the formal counterparts of premises and conclusions. In this respect, there is no substantial difference between the concept(s) of validity proposed by IEL and those proposed by other logics: quite a lot has to be assumed in order to decide whether an inference performed in a natural language is valid in view of a given logic.

The final remark is this. Validity of "declarative" inferences, that is, inferences having declarative sentences as premises and conclusions, is usually defined in terms of (logical) entailment. Yet nobody claims that every inference which is not valid in this sense is substantially faulty: there are inductive inferences, analogical inferences, and so forth. Similarly, besides valid erotetic inferences of the first kind there are invalid, but still plausible, inferences of the analysed type. However, so far IEL does not provide an account of them.

7
Erotetic Implication

The concept of evocation enables us to define validity of erotetic inferences which have questions as conclusions, but not as premises, that is, erotetic inferences of the first kind. Erotetic inferences of the second kind have questions among premises and questions as conclusions. In order to present the account of validity of these inferences proposed by IEL we need the concept of *erotetic implication*.

As in Chapter 6, in the general considerations we assume that we deal with a formal language with questions, $\mathcal{L}$. The language is supposed to satisfy the conditions specified in section 5.4 of Chapter 5, and the semantic concepts introduced in Chapters 3 and 4 apply accordingly. For conciseness, we omit the specifications "in $\mathcal{L}$" and "of $\mathcal{L}$".

7.1 Definition of erotetic implication

Erotetic implication is a ternary relation between a question, a (possibly empty) set of d-wffs, and a question.

Definition 7.1 (*Erotetic implication*). *A question Q implies a question Q_1 on the basis of a set of d-wffs X (in symbols: $\mathbf{Im}(Q, X, Q_1)$) iff:*

1. *for each $A \in \mathsf{d}Q : X \cup \{A\} \models \mathsf{d}Q_1$, and*
2. *for each $B \in \mathsf{d}Q_1$ there exists a non-empty proper subset Y of $\mathsf{d}Q$ such that $X \cup \{B\} \models Y$.*

Recall that $\models$ stands for multiple-conclusion entailment, whereas $\mathsf{d}Q$ and $\mathsf{d}Q_1$ are the sets of direct answers to Q and Q_1, respectively. Thus Q implies Q_1 on the basis of X if, first, the set of direct answers to Q_1 is mc-entailed by each set made up of X and a direct answer to Q, and second, each direct answer to Q_1 mc-entails, together with X, some non-empty proper subset of the set of direct answers to Q.

If $\mathbf{Im}(Q, X, Q_1)$, then Q_1 is said to be the *implied question*, Q is the *implying question*, and the elements of X are called *auxiliary d-wffs*.

We have defined erotetic implication in terms of mc-entailment. A reader not adjusted to operating with mc-entailment can get a better comprehension of erotetic implication when the following "translation" is made. First, think of:

1. for each $A \in \mathsf{d}Q : X \cup \{A\} \Vvdash \mathsf{d}Q_1$

in terms of:

1*. for each $A \in Q$: the set $X \cup \{A\}$ entails a disjunction of all the direct answers to Q_1.

Second,

2. for each $B \in \mathsf{d}Q_1$ there exists a non-empty proper subset Y of $\mathsf{d}Q$ such that $X \cup \{B\} \Vvdash Y$

as an equivalent of:

2*. for each $B \in Q_1$ there exists a non-empty proper subset Y of $\mathsf{d}Q$ such that $X \cup \{B\}$ entails a disjunction of all the elements of Y.

In both cases the corresponding disjunction can be finite or infinite, but is still "classical" in the sense that it is true if, and only if at least one disjunct is true.

Speaking in very general terms: question Q implies question Q_1 on the basis of a set of d-wffs X if, and only if, first, the soundness of Q warrants, together with the truth of all the d-wffs in X, the soundness of Q_1, and second, each direct answer to Q, if true, and if all the d-wffs in X are true, warrants that a true direct answers to Q_1 belongs to a specified proper subset of the set of all the direct answers to Q. The peculiarity of erotetic implication lies in its goal-directness: an implied question is semantically grounded in the implying question and, at the same time, facilitates the answering of the implying question. Let us add: facilitates by *narrowing down*. Clause (2) of Definition 7.1 is equivalent to:

2**. each set made up of a direct answer to Q_1 and X narrows down the set of direct answers to Q.

where the concept of narrowing down is understood in the sense specified by Definition 3.17 (see page 35).

7.1.1 Eliminating a direct answer vs. narrowing down the set of direct answers

As we have already observed[1], narrowing down need not equal elimination.

Let us consider the conditions (in both cases it is assumed that $B \in \mathsf{d}Q_1$):

(ND) there exists a non-empty proper subset Y of $\mathsf{d}Q$ such that $X \cup \{B\} \Vvdash Y$,
(EL) $X \cup \{B\}$ eliminates a certain direct answer C to Q.

---

[1] See section 3.3 of Chapter 3.

7.1 Definition of erotetic implication

Conditions (ND) and (EL) are not equivalent. The following example illustrates this.

Example 7.2. Let $Q = ?\{p,q\}$, $X = \emptyset$, and $Q_1 = ?p$. Thus $\mathsf{d}Q = \{p,q\}$ and $\mathsf{d}Q_1 = \{p, \neg p\}$. Clearly, $\neg p$ eliminates p. But $\neg p$ entails neither p nor q, and hence $\neg p$ does not mc-entail any proper subset of $\{p,q\}$.

Let $Q = ?\{q,r,s\}$, $X = \{p \to q, \neg p \to r \vee s\}$, and $Q_1 = ?p$. Now $X \cup \{\neg p\}$ mc-entails the proper subset $\{r,s\}$ of $\mathsf{d}Q$. However, $X \cup \{\neg p\}$ does not eliminate any direct answer to Q. When one gets $\neg p$, it is still possible that q holds.

However, the following is true:

Corollary 7.3. *Let $B \in \mathsf{d}Q_1$. If condition* (EL) *is fulfilled by $X \cup \{B\}$ and, in addition, the following condition holds:*

(RS) $X \models \mathsf{d}Q$

then condition (ND) *is satisfied by $X \cup \{B\}$.*

Proof. Let $\mathsf{P} = \langle \mathsf{T_P}, \mathsf{U_P} \rangle$ be an arbitrary but fixed admissible partition of the considered language. Since $X \models \mathsf{d}Q$, then also $X \cup \{B\} \models \mathsf{d}Q$. Suppose that $X \cup \{B\} \subset \mathsf{T_P}$. Hence $\mathsf{d}Q \cap \mathsf{T_P} \neq \emptyset$. Let C be a direct answer to Q eliminated by $X \cup \{B\}$. Thus $C \in \mathsf{U_P}$. It follows that $(\mathsf{d}Q \setminus \{C\}) \cap \mathsf{T_P} \neq \emptyset$. Therefore $X \cup \{B\} \models \mathsf{d}Q \setminus \{C\}$. But $\mathsf{d}Q \setminus \{C\}$ is a non-empty proper subset of $\mathsf{d}Q$, since each question is supposed to have at least two direct answers. □

The clause $X \models \mathsf{d}Q$ requires the implying question Q to be sound relative to the set of auxiliary d-wff X. This requirement is neither included in nor implied by Definition 7.1.

We also have:

Corollary 7.4. *Let $B \in \mathsf{d}Q_1$. If condition* (ND) *is fulfilled by $X \cup \{B\}$ and the following is true:*

(ME) *for any $A, C \in \mathsf{d}Q$, where $A \neq C$: A eliminates C*

then condition (EL) *is satisfied by $X \cup \{B\}$.*

Proof. Let Y be a non-empty proper subset of $\mathsf{d}Q$ such that $X \cup \{B\} \models Y$. Suppose that $X \cup \{B\}$ does not eliminate any direct answer to Q. Thus for each $A \in \mathsf{d}Q$ there exists an admissible partition P such that $X \cup \{B, A\} \subset \mathsf{T_P}$. But Y is a proper subset of $\mathsf{d}Q$. Therefore $\mathsf{d}Q \setminus Y \neq \emptyset$. Let $C \in \mathsf{d}Q \setminus Y$ and let P^* be an admissible partition such that $X \cup \{B, C\} \subset \mathsf{T_{P^*}}$. By the condition (ME), C eliminates any other direct answer to Q. Hence $Y \subset \mathsf{U_{P^*}}$. Thus $X \cup \{B\} \not\models Y$. A contradiction. □

Clause (ME) expresses the idea of mutual exclusiveness of direct answers. However, IEL does not assume that direct answers to all questions should be mutually exclusive. But when we consider an implying question, Q, whose direct answers are mutually exclusive, the second clause of the definition of erotetic implication holds just in case the following condition is satisfied:

2^{***}. *for each $B \in \mathsf{d}Q_1$: $X \cup \{B\}$ eliminates a certain direct answer to Q.*

7.1.2 Narrowing down vs. answering

Sometimes narrowing down reduces to answering. Let us consider the following condition:

(DP) $\quad X \cup \{B\}$ entails a direct or partial answer to Q.

Clearly, we have:

Corollary 7.5. *Let $B \in \mathsf{d}Q_1$. If condition* (DP) *is fulfilled by $X \cup \{B\}$, then condition* (ND) *is satisfied by $X \cup \{B\}$.*

Thus the second clause of Definition 7.1 of erotetic implication is fulfilled if each direct answer to the implied question entails, together with the auxiliary d-wffs, some direct or partial answer to the implying question.

There are cases in which narrowing down amounts to answering.

The relation $\models$ of mc-entailment is said to be *compact* if whenever $X \models Y$, then $X_1 \models Y_1$ for some finite subsets X_1 of X, and Y_1 of Y. We say that a language contains disjunction, $\vee$, construed classically just in case the following condition holds:

($CL^\vee$) $\quad$ for each admissible partition $\mathsf{P} = \langle \mathsf{T_P}, \mathsf{U_P} \rangle$ of the language:

$$\{A_1, \ldots, A_n\} \cap \mathsf{T_P} \neq \emptyset \quad \textit{iff} \quad \ulcorner A_1 \vee \ldots \vee A_n \urcorner \in \mathsf{T_P}.$$

One can prove:

Corollary 7.6. *Let $B \in \mathsf{d}Q_1$. If a language includes disjunction construed classically, mc-entailment in the language is compact, and condition* (DP) *holds for $X \cup \{B\}$, then condition* (ND) *is fulfilled by $X \cup \{B\}$.*

Proof. By compactness, $X \cup \{B\} \models Y$ yields $X \cup \{B\} \models Y_1$ for some finite subset Y_1 of Y. Suppose that $Y_1 = \emptyset$. Then $X \cup \{B\} \models \{A\}$ for any $A \in \mathsf{d}Q$. Now suppose that $Y_1 \neq \emptyset$. By ($CL^\vee$) we get $X \cup \{B\} \models C_1 \vee \ldots \vee C_k$, where $Y_1 = \{C_1, \ldots, C_k\}$. If $k = 1$, then $C_1 \in \mathsf{d}Q$; otherwise $C_1 \vee \ldots \vee C_k$ is a partial answer to Q. $\square$

7.1.3 Some comparisons

The concept of erotetic implication was introduced in Wiśniewski (1990a). Without going into historical details[2] let us only compare it with three, in a sense, alternative proposals. The first one comes from Belnap and Steel (1976), the second from Groenendijk and Stokhof (1997). The last one was put forward by Grobler (2006).

---

[2] A brief overview can be found in section 1.6 of Wiśniewski (1995).

7.1 Definition of erotetic implication

Belnap: entailment between quasiformulas

Belnap assigns the logical values, Truth and Falsity, to questions. A question Q is said to be *true in a model* M if at least one direct answer to Q is true in M (for simplicity, we disregard here Belnap's distinction between interrogatives and questions). Let us use the term *quasiformulas* as a cover term for d-wffs and questions of a formal language. A straightforward generalization of the standard concept of entailment is: a set of quasiformulas Γ entails a quasiformula γ iff γ is true in each model in which all the quasiformulas in Γ are true.

As a special case we get: a question Q_1 is entailed by a question Q together with a set of d-wffs X iff $X \cup \{Q\}$ entails Q_1. Given some obvious assumptions, Belnap-style entailment of Q_1 from $X \cup \{Q\}$ holds just in case clause (1) of Definition 7.1 is fulfilled. Yet, recall that erotetic implication is defined by means of *two* clauses.

Groenendijk and Stokhof: interrogative entailment

Let us now consider the proposal present in Groenendijk and Stokhof (1997). The underlying idea is:

> "(...) interrogatives $?\phi_1 \ldots ?\phi_n$ entail (...) interrogative $?\psi$ in a model M iff any proposition which completely answers all of the $?\phi_1 \ldots ?\phi_n$ in M, also completely answers $?\psi$ in M. Logical entailment amounts to entailment in all models." (Groenendijk and Stokhof (1997), p. 1090).

Thus as a special case we get something like: a question Q entails a question Q_1 iff for each model M, any proposition that completely answers Q in M, completely answers Q_1 in M as well.

The idea resembles that of *containment* in the sense of Hamblin (1958): a question Q contains a question Q_1 iff from each answer to Q one can deduce some answer to Q_1.

Clause (2) of Definition 7.1 of erotetic implication expresses an idea similar to that of Groenendijk and Stokhof. However, in our case it is the implied question which facilitates answering the implying question. Moreover, as corollaries 7.5 and 7.6 illustrate, partial answers are allowed. Yet, the Groenendijk-Stokhof's proposal does not provide any counterpart of the first clause of the definition of erotetic implication. On the other hand, if each question considered is supposed to be safe, the condition is trivially fulfilled.

Grobler: falsificationist erotetic implication

Grobler (cf. Grobler (2006), Grobler (2012)) defines a ternary relation between a question, a set of sentences, and a question, which he coins *falsificationist erotetic implication*. By using the conceptual apparatus of this book, Grobler's notion can be defined as follows:

Definition 7.7 (*Falsificationist erotetic implication*). A question Q f-implies a question Q_1 on the basis of a set of d-wffs X (in symbols: $\mathbf{Im_f}(Q, X, Q_1)$) iff:

1. for each $A \in \mathsf{d}Q : X \cup \{A\} \models \mathsf{d}Q_1$, and
2. for some $B \in \mathsf{d}Q_1 \colon X \cup \{B\}$ eliminates a direct answer to Q.

The scopes of **Im** and **Im$_\mathsf{f}$** overlap, but neither is included in the other. By Corollary 7.4, **Im** yields **Im$_\mathsf{f}$** for implying questions with mutually exclusive direct answers. On the other hand, clause (2) of Definition 7.7 is only existential, while the second clause of Definition 7.1 of erotetic implication speaks about each direct answer to Q_1. If we strengthened clause (2) of the above definition to the general one (i.e. to the clause (2***) discussed in section 7.1.1 above), the "strong" falsificationist erotetic implication defined that way would yield IEL's erotetic implication given that the condition (RS) (again, see section 7.1.1) is satisfied by X and Q, that is, the implying question is sound relative to the auxiliary d-wffs.

Grobler's erotetic implication, nevertheless, has substantial applications in the areas of philosophy of science and epistemology.[3]

7.2 Erotetic implication and validity

The clauses (1) and (2) of Definition 7.1 of erotetic implication express in exact terms the intuitions which lie behind the conditions of validity ($\mathbf{C_3}$) and ($\mathbf{C_4}$) pertaining to erotetic inferences of the second kind.[4] Let us recall them:

($\mathbf{C_3}$) (TRANSMISSION OF SOUNDNESS/TRUTH INTO SOUNDNESS). *If the initial question is sound and all the declarative premises are true, then the question which is the conclusion must be sound.*

($\mathbf{C_4}$) (OPEN-MINDED COGNITIVE USEFULNESS). *For each direct answer B to the question which is the conclusion there exists a non-empty proper subset Y of the set of direct answers to the initial question such that the following condition holds:*
(♣) *if B is true and all the declarative premises are true, then at least one direct answer $A \in Y$ to the initial question must be true.*

We have:

Corollary 7.8. *The condition:*

(1) for each $A \in \mathsf{d}Q : X \cup \{A\} \models \mathsf{d}Q_1$

is fulfilled iff the following condition holds:

(1') for each admissible partition $\mathsf{P} = \langle \mathsf{T_P}, \mathsf{U_P} \rangle$ of the language: if Q is sound in P and $X \subset \mathsf{T_P}$, then Q_1 is sound in P.

Proof. Assume that (1) holds and that (1') does not hold. Thus there exists an admissible partition, P, such that $\mathsf{d}Q_1 \subset \mathsf{U_P}$, $X \subset \mathsf{T_P}$, and $A \in \mathsf{T_P}$ for some $A \in \mathsf{d}Q$. Hence $X \cup \{A\} \not\models \mathsf{d}Q_1$. A contradiction.

---
[3] See Grobler (2006), Grobler (2012).
[4] See section 5.2.2 of Chapter 5.

Assume that (1') holds and (1) does not hold. Hence for some $A \in dQ$ we have $X \cup \{A\} \not\models dQ_1$. It follows that there exists an admissible partition, P, such that Q is sound in P, $X \subset \mathsf{T_P}$, and Q_1 is not sound in P. A contradiction again. □

Corollary 7.9. *The condition:*

(2) for each $B \in dQ_1$ there exists a non-empty proper subset Y of dQ such that $X \cup \{B\} \models Y$

is satisfied iff the following condition is fulfilled:

(2') for each $B \in dQ_1$ there exists a non-empty proper subset Y of dQ such that for each admissible partition $\mathsf{P} = \langle \mathsf{T_P}, \mathsf{U_P} \rangle$ of the language: if $X \cup \{B\} \subset \mathsf{T_P}$, then $A \in \mathsf{T_P}$ for some $A \in Y$.

Proof. It suffices to observe that, by the definition of mc-entailment, $X \cup \{B\} \models Y$ just in case for each admissible partition P: if $X \cup \{B\} \subset \mathsf{T_P}$, then $A \in \mathsf{T_P}$ for some $A \in Y$. □

The modal element present in the conditions (**C₃**) and (**C₄**) is mirrored by the reference to the class of admissible partitions.

Validity of erotetic inferences of the second kind, that is, inferences leading from questions, on the basis of sets of declarative sentences, to questions, is then defined as follows.

Definition 7.10 (*Validity II*).
An erotetic inference of the second kind, $\langle Q, X, Q_1 \rangle$, is valid iff $\mathbf{Im}(Q, X, Q_1)$.

As previously, by proposing the above definition we implicitly assume that the X's, as well as Q and Q_1, are expressions of a language for which erotetic implication is appropriately defined. Again, since erotetic implication has been defined for formal languages, the above concept of validity can be applied to erotetic inferences performed in natural languages only in an indirect way: an inference is valid just in case erotetic implication holds between the formal counterparts of premises and conclusions.

Comments. Let us recall: validity is a normative concept and in the case of inferences having questions as conclusions the notion of validity is neither given by God nor by Tradition. As long as erotetic inferences of the second kind are concerned, IEL proposes to construe their validity according to the ideas expressed by the conditions (**C₃**) and (**C₄**), and hence defines validity in terms of erotetic implication. While condition (**C₃**) is relatively unquestionable, the acceptance of condition (**C₄**) may raise doubts due to its strength. The condition, and erotetic implication thereof, require *every* direct answer to the question which is the conclusion to be potentially cognitively useful. One can propose a weakening of this general requirement or/and a reconsideration of the underlying (semantic) concept of cognitive usefulness. From the standpoint of IEL, however, an outcome of such an analysis amounts to a characterization of a class of erotetic inferences of the second kind which need not be IEL-valid, but, nevertheless, are worth to be distinguished. In particular, Grobler's proposal (see section 7.1.3 above) can also be applied this way.

7.3 Some properties of erotetic implication

For the properties of erotetic implication see Wiśniewski (1990a), Wiśnie-wski (1994a), Wiśniewski (1995), Wiśniewski (1996), and Wiśniewski (2001). Below we will describe only some of them, mainly those which shed light on some interesting features of valid erotetic inferences of the second kind.

Till the end of this section by "inferences", unless otherwise stated, we will mean erotetic inferences of the second kind.

7.3.1 Mutual soundness

The following is true:

Corollary 7.11. *Let* $\mathbf{Im}(Q, X, Q_1)$. *Then* $X \models dQ$ *iff* $X \models dQ_1$.

Proof. Let $X \models dQ$. Suppose that $X \not\models dQ_1$. Thus for some admissible partition P we have $X \subset \mathsf{T}_\mathsf{P}$ and $dQ_1 \subset \mathsf{U}_\mathsf{P}$. Hence, by clause (1) of Definition 7.1, $A \in \mathsf{U}_\mathsf{P}$ for each $A \in dQ$ and therefore $dQ \subset \mathsf{U}_\mathsf{P}$. It follows that $X \not\models dQ$. A contradiction.

Now suppose that $X \models dQ_1$, but $X \not\models dQ$. Thus there exists an admissible partition, P, such that $X \subset \mathsf{T}_\mathsf{P}$ and $dQ \subset \mathsf{U}_\mathsf{P}$. Hence $Y \subset \mathsf{U}_\mathsf{P}$ for any $Y \subset dQ$. By clause (2) of Definition 7.1 we get $B \in \mathsf{U}_\mathsf{P}$ for any $B \in dQ_1$, and thus $dQ_1 \subset \mathsf{U}_\mathsf{P}$. Hence $X \not\models dQ_1$. A contradiction again. □

A question being a premise of a valid inference need not be sound relative to the declarative premises. Corollary 7.11 shows, however, that the "question-premise" is sound relative to the declarative premises if, and only if the question which is the conclusion is sound relative to them.

As an immediate consequence of Corollary 7.11 we get:

Corollary 7.12. *Let* $\mathbf{Im}(Q, X, Q_1)$ *and let* $\mathsf{P} = \langle \mathsf{T}_\mathsf{P}, \mathsf{U}_\mathsf{P} \rangle$ *be an admissible partition of the language such that* $X \subset \mathsf{T}_\mathsf{P}$. *Then* Q_1 *is sound in* P *iff* Q *is sound in* P.

Tables 7.1 and 7.2 display possible connections. We assume that Q (erotetically) implies Q_1 on the basis of X, and that $\mathsf{P} = \langle \mathsf{T}_\mathsf{P}, \mathsf{U}_\mathsf{P} \rangle$ is an arbitrary but fixed admissible partition of the relevant language.

Table 7.1. *From implying question to implied question.*

Q	X	Q_1
sound in P	$X \subset \mathsf{T}_\mathsf{P}$	sound in P
unsound in P	$X \subset \mathsf{T}_\mathsf{P}$	unsound in P
sound in P	$X \not\subset \mathsf{T}_\mathsf{P}$	sound in P or unsound in P
unsound in P	$X \not\subset \mathsf{T}_\mathsf{P}$	sound in P or unsound in P

What comes as a surprise are: the second row of table 7.1, and the first row of table 7.2. It is clear that a valid inference has to lead from a sound question to a sound question given that the declarative premises are true. But the relevant

Table 7.2. *From implied question to implying question.*

Q_1	X	Q
sound in P	$X \subset \mathsf{T_P}$	sound in P
unsound in P	$X \subset \mathsf{T_P}$	unsound in P
sound in P	$X \not\subset \mathsf{T_P}$	sound in P or unsound in P
unsound in P	$X \not\subset \mathsf{T_P}$	sound in P or unsound in P

rows show that validity defined in terms of erotetic implication warrants even more. The second row of table 7.1 shows that a valid inference based on true declarative premises, but unsound erotetic premise always leads to an unsound conclusion. Thus one cannot pass, in a valid inference and with the help of true declarative premises, from a question that does not have a true direct answer to a question which has such answer. The first row of table 7.2 shows, in turn, that a valid inference with sound conclusion, based on true declarative premises, always has sound erotetic premise. Thus if a questioner has passed, in a valid inference and with the help of true declarative premises, to a question which has a true direct answer, this confirms that his/her initial question has a true direct answer as well. Both warranties are substantial, since (as in the case of inferences with declarative premises and conclusions) the semantic status, i.e. soundness/unsoundness, of a question being a premise or a conclusion can be either unknown to a questioner or inadequately assessed by him/her.

7.3.2 Monotony and transitivity issues

It is obvious that erotetic implication is monotone with respect to d-wffs. We have:

Corollary 7.13. *If* $\mathbf{Im}(Q, X, Q_1)$ *and* $X \subset Y$, *then* $\mathbf{Im}(Q, Y, Q_1)$.

Thus when we extend the set of declarative premises of a valid erotetic inference of the second kind, validity is still retained.

Question $Q^\star$ is an *extension* of question Q if $\mathsf{d}Q$ is a proper subset of $\mathsf{d}Q^\star$. A moment's reflection shows that the following hold:

Corollary 7.14. *Let* $\mathbf{Im}(Q, X, Q_1)$, *and let* $Q^\star$ *be an extension of* Q, *Then* $\mathbf{Im}(Q^\star, X, Q_1)$ *iff* $X \cup C \models \mathsf{d}Q_1$ *for each* $C \in \mathsf{d}Q^\star \setminus \mathsf{d}Q$.

Corollary 7.15. *Let* $\mathbf{Im}(Q, X, Q_1)$, *and let* $Q_1^\star$ *be an extension of* Q_1. *Then* $\mathbf{Im}(Q, X, Q_1^\star)$ *iff for each* $C \in \mathsf{d}Q_1^\star \setminus \mathsf{d}Q_1$ *there exists a non-empty proper subset* Y *of* $\mathsf{d}Q$ *such that* $X \cup \{C\} \models Y$.

Thus when we extend an implying question or an implied question, validity need not be retained, but is retained given that some additional conditions (specified above) are met.

Erotetic implication is not "transitive" in the sense that there are cases in which $\mathbf{Im}(Q, X, Q_1)$ and $\mathbf{Im}(Q_1, X, Q_2)$ hold, but $\mathbf{Im}(Q, X, Q_2)$ does not hold. Here is a simple (counter)example taken from language $\mathcal{L}_{cpl}^?$. We have:

$$\mathbf{Im}(?p, \emptyset, ?\{p \wedge q, p \wedge \neg q, \neg p\}) \tag{7.1}$$

$$\mathbf{Im}(?\{p \wedge q, p \wedge \neg q, \neg p\}, \emptyset, ?q) \tag{7.2}$$

but we (fortunately!) do not have $\mathbf{Im}(?p, \emptyset, ?q)$. However, some special kinds of erotetic implication are "transitive" (see below).

7.4 Some special kinds of erotetic implication

7.4.1 Regular erotetic implication

The second clause of Definition 7.1 of erotetic implication can be fulfilled in such a way that the relevant Y's are singleton sets. In this case we get regular erotetic implication.

Definition 7.16 (*Regular erotetic implication*). *A question Q regularly implies a question Q_1 on the basis of a set of d-wffs X iff*

1. *for each $A \in \mathsf{d}Q : X \cup \{A\} \models \mathsf{d}Q_1$, and*
2. *for each $B \in \mathsf{d}Q_1$ there exists $C \in \mathsf{d}Q$ such that $X \cup \{B\} \models C$.*

Regular erotetic implication is "transitive" in the sense explained by:

Corollary 7.17. *If Q regularly implies Q_1 on the basis of X, and Q_1 regularly implies Q_2 on the basis of X, then Q regularly implies Q_2 on the basis of X.*

7.4.2 Strong erotetic implication

The second clause of the definition of erotetic implication can be trivially satisfied due to the fact that X alone mc-entails some non-empty proper subset of $\mathsf{d}Q$. This cannot happen, however, in the case of the so-called strong erotetic implication.

Definition 7.18 (*Strong erotetic implication*). *A question Q strongly implies a question Q_1 on the basis of a set of d-wffs X iff*

1. *for each $A \in \mathsf{d}Q : X \cup \{A\} \models \mathsf{d}Q_1$, and*
2. *for each $B \in \mathsf{d}Q_1$ there exists a non-empty proper subset Y of $\mathsf{d}Q$ such that $X \cup \{B\} \models Y$, but $X \not\models Y$.*

A regular strong erotetic implication is a special case; its definition is straightforward.

7.4.3 Pure erotetic implication. Analyticity

Pure erotetic implication is erotetic implication on the basis of the empty set.

Definition 7.19 (*Pure erotetic implication*). *A question Q implies a question Q_1 (in symbols: $\mathbf{Im}(Q, Q_1)$) iff:*

1. *for each $A \in \mathsf{d}Q : A \models \mathsf{d}Q_1$, and*

2. for each $B \in \mathsf{d}Q_1$ there exists a non-empty proper subset Y of $\mathsf{d}Q$ such that $B \mathrel{\|\!\!=} Y$.

Regular pure erotetic implication can be defined accordingly.

Recall that $?A$ abbreviates $?\{A, \neg A\}$. For further reference let us note:

Corollary 7.20. *If $A \models B$ and $B \models A$, then $\mathbf{Im}(?A, ?B)$.*

For languages with negation we can define another special kind of pure erotetic implication: the analytic one.

Definition 7.21 (Analytic erotetic implication). *A question Q analytically implies a question Q_1 iff $\mathbf{Im}(Q, Q_1)$ and each immediate subformula of a direct answer to Q_1 is a subformula of a direct answer to Q or is a negation of a subformula of a direct answer to Q.*

Erotetic inferences which do not involve any declarative premise(s), but have questions as premises and conclusions are instances of erotetic inferences of the second kind. The following corollary characterizes an important feature of valid erotetic inferences in which no declarative premise occurs.

Corollary 7.22. *Let $\mathbf{Im}(Q, Q_1)$. Then Q_1 is safe iff Q is safe, and Q_1 is risky iff Q is risky.*

Proof. By Corollary 4.15 and Corollary 7.11. □

Thus a valid erotetic inference of the analysed kind always leads from a safe question to a safe question, and from a risky question to a risky question. Moreover, one cannot arrive at a safe question when the premise is not a safe question, and similarly for riskiness.

7.5 Examples of erotetic implication

As in the case of evocation, Definition 7.1 of erotetic implication is schematic: when a language is specified, one gets the definition of erotetic implication in the language. To indicate that erotetic implication in a given language is considered, one can add a subscript to $\mathbf{Im}$.

Let us now present some examples of erotetic implication in the languages $\mathcal{L}^?_{cpl}$ and $\mathcal{L}^?_{fom}$. In what follows we use the letters A, B, C, D, with subscripts if needed, as metalanguage variables for d-wffs of $\mathcal{L}^?_{cpl}$ and $\mathcal{L}^?_{fom}$. When no quantifiers or symbols referring to individual variables or constants occur, the formulas presented below refer both to erotetic implication in $\mathcal{L}^?_{cpl}$ and $\mathcal{L}^?_{fom}$; otherwise only $\mathcal{L}^?_{fom}$ is taken into consideration. We shall supplement the examples with comments included in square brackets, which will indicate the kind of erotetic implication involved. The comment "not regular" means "it is not the case that each instance is regular", and similarly for analyticity.

7.5.1 Pure erotetic implication: examples

Recall that $?A$ stands for a simple yes-no question to be read "Is it the case that A?". The following:
$$? \pm |A, B|$$
abbreviates:
$$?\{A \wedge B, A \wedge \neg B, \neg A \wedge B, \neg A \wedge \neg B\}$$
which can be read "Is it the case that A and is it the case that B?"; the direct answers are: $A \wedge B$, $A \wedge \neg B$, $\neg A \wedge B$, $\neg A \wedge \neg B$. A question of the form:
$$?\{A \wedge B, A \wedge \neg B, \neg A\}$$
is a conditional question with revocable antecedent and can be read "Is it the case that A?; if so, is it also the case that B?".

As long as $\mathcal{L}^?_{fom}$ is concerned, the relevant metalanguage variables that occur in whether-questions are supposed to represent sentences (i.e. closed d-wffs). Metalinguistic expressions of the form Ax and Bx refer to sentential functions of $\mathcal{L}^?_{fom}$ with only one free variable. Recall that the sentences that occur in a whether-question are supposed to be pairwise syntactically distinct.

Here are examples of pure erotetic implication.

$$\mathbf{Im}(?\neg A, ?A) \qquad (7.3)$$
$$[regular,\ analytic]$$

$$\mathbf{Im}(?A, ?\neg A) \qquad (7.4)$$
$$[regular,\ analytic]$$

$$\mathbf{Im}(?\{A, B\}, ?\{B, A\}) \qquad (7.5)$$
$$[regular,\ analytic]$$

$$\mathbf{Im}(?\{A, B \vee C\}, ?\{A, B, C\}) \qquad (7.6)$$
$$[regular,\ analytic]$$

$$\mathbf{Im}(?\{A, B, C\}, ?\{A, B \vee C\}) \qquad (7.7)$$
$$[not\ regular,\ analytic]$$

Note that although in the cases (7.6) and (7.7) the implying question and the implied question switch their places, this is the regularity condition which makes the difference. As we will show below, sometimes analyticity performs an analogous role.

$$\mathbf{Im}(? \pm |A, B|, ?A) \qquad (7.8)$$
$$[not\ regular,\ analytic]$$

$$\mathbf{Im}(? \pm |A, B|, ?B) \qquad (7.9)$$
$$[not\ regular,\ analytic]$$

$$\mathbf{Im}(?A, ? \pm |A, B|) \qquad (7.10)$$
$$[regular,\ not\ analytic]$$

$$\mathbf{Im}(?B, ? \pm |A, B|) \qquad (7.11)$$
$$[regular,\ not\ analytic]$$

7.5 Examples of erotetic implication

$$\mathbf{Im}(?\pm|A,B|, ?(A\otimes B)) \qquad (7.12)$$
[*not regular, not analytic if* $\otimes \neq \wedge$]

where $\otimes$ is any of the connectives: $\wedge, \vee, \rightarrow, \leftrightarrow$.

$$\mathbf{Im}(?(A\otimes B), ?\pm|A,B|) \qquad (7.13)$$
[*regular, analytic*]

where $\otimes$ is any of the connectives: $\wedge, \vee, \rightarrow, \leftrightarrow$.

$$\mathbf{Im}(?\{A\wedge B, A\wedge \neg B, \neg A\}, ?A) \qquad (7.14)$$
[*not regular, analytic*]

$$\mathbf{Im}(?A, ?\{A\wedge B, A\wedge \neg B, \neg A\}) \qquad (7.15)$$
[*regular, not analytic*]

$$\mathbf{Im}(?\{A, \neg A, B\}, ?B) \qquad (7.16)$$
[*not regular, analytic*]

$$\mathbf{Im}(?\{A, \neg A, \neg B\}, ?B) \qquad (7.17)$$
[*not regular, analytic*]

$$\mathbf{Im}(?\forall x Ax, ?\exists x\neg Ax) \qquad (7.18)$$
[*regular*]

$$\mathbf{Im}(?\exists x Ax, ?\forall x\neg Ax) \qquad (7.19)$$
[*regular*]

$$\mathbf{Im}(?\forall x Ax, ?\pm|\exists x Ax, \exists x\neg Ax|) \qquad (7.20)$$
[*regular*]

$$\mathbf{Im}(?\pm|\exists x Ax, \exists x\neg Ax|, ?\exists x Ax) \qquad (7.21)$$
[*not regular*]

$$\mathbf{Im}(?\exists x Ax, ?\pm|\forall x Ax, \forall x\neg Ax|) \qquad (7.22)$$
[*regular*]

$$\mathbf{Im}(?\pm|\forall x Ax, \forall x\neg Ax|, ?\forall x Ax) \qquad (7.23)$$
[*not regular*]

$$\mathbf{Im}(?\forall x Ax, ?\{\exists x\neg Ax, \neg\exists x\neg Ax, \neg\exists x Ax\}) \qquad (7.24)$$
[*regular*]

$$\mathbf{Im}(?\exists x Ax, ?\{\forall x\neg Ax, \neg\forall x\neg Ax, \forall x Ax\}) \qquad (7.25)$$
[*regular*]

One can easily get further examples of pure erotetic implication by applying Corollary 7.20.

7.5.2 Erotetic implication on the basis of non-empty sets of d-wffs: examples

For brevity, we simply list the elements of the relevant sets of auxiliary d-wffs. The following hold:

$$\mathbf{Im}(?A, A \leftrightarrow B, ?B) \tag{7.26}$$
[*regular*]

$$\mathbf{Im}(?A, B \to A, C \to \neg A, B \lor C, ?\{B, C\}) \tag{7.27}$$
[*regular*]

$$\mathbf{Im}(?(A \otimes B), A^\circ, ?B) \tag{7.28}$$
[*regular*]

where $\otimes$ is any of the connectives: $\land, \lor, \to, \leftrightarrow$, and A° equals $\neg A$ or equals A.

$$\mathbf{Im}(?(A \otimes B), B^\circ, ?A) \tag{7.29}$$
[*regular*]

where $\otimes$ is any of the connectives: $\land, \lor, \to, \leftrightarrow$, and B° equals $\neg B$ or equals B.

$$\mathbf{Im}(?A, B \to A, ?\{A, \neg A, B\}) \tag{7.30}$$
[*regular*]

$$\mathbf{Im}(?A, A \to B, ?\{A, \neg A, \neg B\}) \tag{7.31}$$
[*regular*]

A digression. It is not the case that $?A$ erotetically implies $?B$ on the basis of $B \to A$. One can consider this as a shortcoming. But $?B$ is accessible from $?A$ and $B \to A$ in *two steps*, by applying (7.30) and then (7.16):

$$\mathbf{Im}(?A, B \to A, ?\{A, \neg A, B\})$$
$$\mathbf{Im}(?\{A, \neg A, B\}, ?B)$$

Each of the steps in a valid erotetic inference.

Similarly, $?B$ is accessible from $?A$ and $A \to B$ in *two steps*, by (7.31) and (7.17):

$$\mathbf{Im}(?A, A \to B, ?\{A, \neg A, \neg B\})$$
$$\mathbf{Im}(?\{A, \neg A, \neg B\}, ?B)$$

Transitions from $?\forall x Ax$ to $?\exists x Ax$, and from $?\exists x Ax$ to $?\forall x Ax$ also require two steps. In the former case we apply (7.24) and a special case of (7.17):[5]

$$\mathbf{Im}(?\forall x Ax, ?\{\exists x \neg Ax, \neg \exists x \neg Ax, \neg \exists x Ax\})$$
$$\mathbf{Im}(?\{\exists x \neg Ax, \neg \exists x \neg Ax, \neg \exists x Ax\}, ?\exists x Ax)$$

whereas in the latter case we rely on (7.25) and an instance of (7.16):[6]

$$\mathbf{Im}(?\exists x Ax, ?\{\forall x \neg Ax, \neg \forall x \neg Ax, \forall x Ax\})$$

---

[5] Alternatively, (7.20) and (7.21).
[6] Or on (7.22) and (7.23).

$$\mathbf{Im}(?\{\forall x \neg Ax, \neg \forall x \neg Ax, \forall x Ax\}, ?\forall x Ax)$$

As in the case of examples of evocation, $c, c_1, c_2, \ldots$ are metalanguage variables for individual constants; it is assumed that distinct metalanguage variables represent distinct individual constants.[7] We have:

$$\mathbf{Im}(?A(x/c), \forall x(Ax \leftrightarrow Bx), ?B(x/c)) \qquad (7.32)$$
$$[regular]$$

$$\mathbf{Im}(?A(x/c), \forall x(Ax \leftrightarrow Bx \wedge Cx), ? \pm |B(x/c), C(x/c)|) \qquad (7.33)$$
$$[regular]$$

$$\mathbf{Im}(?\{A(x/c_1), A(x/c_2)\}, \forall x(Ax \leftrightarrow Bx \wedge Cx),$$
$$B(x/c_1), C(x/c_2), ?\{B(x/c_2), C(x/c_1)\}) \qquad (7.34)$$
$$[regular]$$

$$\mathbf{Im}(?\{A(x/c_1), \ldots, A(x/c_n)\}, \exists x(Ax \wedge (x = c_1 \vee \ldots \vee x = c_n)), ?A(x/c_i)) \quad (7.35)$$
$$[regular \text{ if } n = 2]$$

where $n > 1$ and $1 \leq i \leq n$.

A notational convention. We abbreviate

$$?\{A_1, \ldots, A_n\}$$

as:
$$?[A_{|n}] \qquad (7.36)$$

We have:
$$\mathbf{Im}(?[A_{|n}], A_1 \vee \ldots \vee A_n, ?A_i) \qquad (7.37)$$
$$[regular \text{ if } n = 2]$$

where $n > 1$ and $1 \leq i \leq n$.

A digression. Observe that we *do not* have $\mathbf{Im}(?[A_{|n}], ?A_i)$. The reason is that $\neg A_i$ need not mc-entail a proper subset of $\{A_1, \ldots, A_n\}$. But $\{A_1 \vee \ldots \vee A_n, \neg A_i\}$ mc-entails $\{A_1, \ldots, A_n\} \setminus \{A_i\}$.

The following hold:

$$\mathbf{Im}(?[A_{|n}], A_1 \vee \ldots \vee A_n, \neg(A_1 \wedge \ldots \wedge A_n), ?[\neg A_{|n}]) \qquad (7.38)$$
$$[regular \text{ if } n = 2]$$

where $n > 1$.

$$\mathbf{Im}(?[A_{|n}], B_1 \leftrightarrow A_1, \ldots, B_n \leftrightarrow A_n, ?[B_{|n}]) \qquad (7.39)$$
$$[regular]$$

where $n > 1$.

---

[7] When double indices are used, this assumption is, unless otherwise stated, cancelled; cf. (7.56) below.

$$\mathbf{Im}(?[A_{|n}], B_1 \to A_1, \ldots, B_n \to A_n, B_1 \vee \ldots \vee B_n, ?[B_{|n}]) \quad (7.40)$$
$$[\textit{regular}]$$

where $n > 1$.

$$\mathbf{Im}(?[A_{|n}], B \to A_1 \vee \ldots \vee A_{i-1}, \neg B \to A_i \vee \ldots \vee A_n, ?B) \quad (7.41)$$
$$[\textit{regular if } n = 2]$$

where $n > 1$ and $1 < i \leq n$.

$$\mathbf{Im}(?[A_{|n}], B \to A_1 \vee \ldots \vee A_{i-1}, C \to A_i \vee \ldots \vee A_n, B \vee C, ?\{B, C\}) \quad (7.42)$$
$$[\textit{regular if } n = 2]$$

where $n > 1$ and $1 < i \leq n$.

$$\mathbf{Im}(?[A_{|n}], B, ?\{B \to A_1, \ldots, B \to A_n\}) \quad (7.43)$$
$$[\textit{regular}]$$

where $n > 1$.

Recall that an existential which-question, $?S(Ax)$, can be read "Which x is such that Ax?". Recall also that admissible partitions of $\mathcal{L}^?_{fom}$ are determined by those models in which each element of the domain is "named" by some individual constant(s). As a result, we have:

$$\exists x Ax \models_{\mathcal{L}^?_{fom}} \mathbf{S}(Ax) \quad (7.44)$$

and, for any individual constant c:

$$\{\exists x Ax, \neg A(x/c)\} \models_{\mathcal{L}^?_{fom}} (\mathbf{S}(Ax) \setminus \{A(x/c)\}) \quad (7.45)$$

$$\mathbf{Im}(?\mathbf{S}(Ax), \exists x Ax, ?A(x/c)) \quad (7.46)$$
$$[\textit{not regular}]$$

$$\mathbf{Im}(?\mathbf{S}(Ax), \exists x Ax, \forall x(Ax \leftrightarrow Bx), ?B(x/c)) \quad (7.47)$$
$$[\textit{not regular}]$$

$$\mathbf{Im}(?\mathbf{S}(Ax), \forall x(Ax \leftrightarrow Bx), ?\mathbf{S}(Bx)) \quad (7.48)$$
$$[\textit{regular}]$$

$$\mathbf{Im}(?\mathbf{S}(Ax), B \to \exists x(Ax \wedge (x = c_1 \vee \ldots \vee x = c_{i-1})),$$
$$C \to \exists x(Ax \wedge (x = c_i \vee \ldots \vee x = c_n)), B \vee C, ?\{B, C\}) \quad (7.49)$$
$$[\textit{regular if } n = 2]$$

where $n > 1$ and $1 < i \leq n$.

$$\mathbf{Im}(?\mathbf{S}(Ax), \forall x(Bx \to Ax), \exists x(Bx \wedge (x = c_1 \vee \ldots \vee x = c_n)),$$
$$?\{B(x/c_1), \ldots, B(x/c_n)\}) \quad (7.50)$$
$$[\textit{regular}]$$

where $n > 1$.

7.5 Examples of erotetic implication

$$\mathbf{Im}(?\mathbf{S}(Ax), \forall x(Ax \to x = c_1 \vee \ldots \vee x = c_n), ?\{A(x/c_1), \ldots, A(x/c_n)\}) \quad (7.51)$$
[regular]

where $n > 1$.

$$\mathbf{Im}(?\mathbf{S}(Ax), \exists x Ax, \forall x(Ax \to x = c_1 \vee \ldots \vee x = c_n), ?A(x/c_i)) \quad (7.52)$$
[regular if $n = 2$]

where $1 \leq i \leq n$.

$$\mathbf{Im}(?\mathbf{S}(Ax), \exists x Bx, \forall x(Bx \to Ax), ?\mathbf{S}(Bx)) \quad (7.53)$$
[regular]

$$\mathbf{Im}(?\mathbf{S}(Ax), \exists x Ax, \neg \forall x Ax, ?\mathbf{S}(\neg Ax)) \quad (7.54)$$
[not regular]

A general which-question, $?\mathbf{U}(Ax)$, can be read "What are all of the x's such that Ax?". Its direct answers fall under the schema:

$$A(x/c_1) \wedge \ldots \wedge A(x/c_n) \wedge \forall x(Ax \to x = c_1 \vee \ldots \vee x = c_n)$$

where $n \geq 1$ and $c_1, \ldots, c_n$ stand for distinct individual constants.

$$\mathbf{Im}(?\mathbf{U}(Ax), \forall x(Ax \leftrightarrow Bx), ?\mathbf{U}(Bx)) \quad (7.55)$$
[regular]

$$\mathbf{Im}(?\mathbf{U}(Ax), B \to \forall x(Ax \leftrightarrow x = c_{i_1} \vee \ldots \vee x = c_{i_n}),$$
$$C \to \forall x(Ax \leftrightarrow x = c_{j_1} \vee \ldots \vee x = c_{j_k}), B \vee C, ?\{B, C\}) \quad (7.56)$$
[regular]

where $n \geq 1$ and $k \geq 1$.

$$\mathbf{Im}(?\mathbf{S}(Ax), \exists_{\geq k} x Ax, ?\mathbf{U}(Ax)) \quad (7.57)$$
[regular]

where $\exists_{\geq k} x$ is the numerical quantifier "there exist at most k", and $k \geq 1$.

$$\mathbf{Im}(?\mathbf{S}(Ax), \forall x(Bx \to Ax), \exists x Bx, \exists_{\geq k} x Bx, ?\mathbf{U}(Bx)) \quad (7.58)$$
[regular]

where $k \geq 1$.

$$\mathbf{Im}(?\mathbf{S}(Ax), \exists x Ax, \forall x(Ax \to \neg Bx), \exists x Bx, \exists_{\geq k} x Bx, ?\mathbf{U}(Bx)) \quad (7.59)$$
[not regular]

where $k \geq 1$.

$$\mathbf{Im}(?\mathbf{U}(Ax), \exists_{\geq k} x Ax, A(x/c_1) \wedge \ldots \wedge A(x/c_n),$$
$$?\forall x(Ax \to x = c_1 \vee \ldots \vee x = c_n)) \quad (7.60)$$
[not regular]

where $n \leq k$ and $k \geq 1$.

For further examples of erotetic implication (in different languages) see e.g. Wiśniewski (1990a), Wiśniewski (1994a), Wiśniewski (1995) (Chapter 7), or Wiśniewski (2001). See also the following chapters of this book.

7.6 Erotetic implication, evocation, and goal-directness

IEL gives an account of validity of erotetic inferences of the first and second kinds. These inferences differ with regard to types of premises involved, and the intuitions which underlie the respective concepts of validity are diverse. However, from a purely formal point of view evocation of questions and erotetic implication are interrelated in many ways. Let us end this chapter by pointing out some of the connections.

7.6.1 Evocation as erotetic implication by non-factual questions

We say that a d-wff A of a language $\mathcal{L}$ is *valid* iff for each admissible partition of the language, A is true in the partition. We need the following auxiliary concept:

Definition 7.23 (Non-factual question). *A question Q of $\mathcal{L}$ is non-factual iff each direct answer to Q is a valid d-wff.*

Since no valid d-wff is a carrier of a factual information, the label "non-factual question" seems appropriate. Needless to say, non-factual questions are safe and self-rhetorical. Of course, there is no warranty that every language of the considered kind includes non-factual question(s).

Recall that a question Q is *informative relative to* a set of d-wffs X iff no direct answer to Q is entailed by X.

The following holds:

Theorem 7.24. *Let $\mathcal{L}$ be a language of the considered kind in which non-factual question(s) occur. Then $\mathbf{E}_\mathcal{L}(X,Q)$ iff*

1. *Q is informative relative to X, and*
2. *for each non-factual question $Q^\star$ of the language: $\mathbf{Im}_\mathcal{L}(Q^\star, X, Q)$.*

Proof. Let $Q^\star$ be a non-factual question.

Suppose that $\mathbf{E}_\mathcal{L}(X,Q)$. Thus Q is informative relative to X, and $X \models_\mathcal{L} dQ$. Clearly $X \cup \{A\} \models_\mathcal{L} dQ$ for any $A \in dQ^\star$, and $X \cup \{B\} \models_\mathcal{L} A$ for any $A \in dQ^\star$ and $B \in dQ$. Hence $\mathbf{Im}_\mathcal{L}(Q^\star, X, Q)$.

Suppose that $\mathbf{Im}_\mathcal{L}(Q^\star, X, Q)$. Since $Q^\star$ is non-factual, $X \models_\mathcal{L} dQ^\star$. Thus, by Corollary 7.11, $X \models_\mathcal{L} dQ$. By assumption, Q is informative relative to X. Therefore $\mathbf{E}_\mathcal{L}(X,Q)$. □

By and large, Theorem 7.24 says that evocation amounts to erotetic implication of informative questions by non-factual questions. Thus it is possible to define evocation (in a somewhat tricky way) in terms of erotetic implication. However, when it comes to the validity of erotetic inferences, a substantial difference still remains. Valid erotetic inferences of the second kind are, in a sense, goal-directed: one derives a question from a question in order to facilitate the answering of the initial question. On the other hand, the conclusion of a valid erotetic inference of the first kind need not be dependent upon previous goals and usually sets a goal by itself.

7.6.2 Answering evoked questions by means of answers to implied questions

Let Ξ be a non-empty set of questions. By a Ξ-*answer set* we mean a set of d-wffs that comprises only direct answers to questions of Ξ and such that the set includes exactly one direct answer to each question of Ξ. For instance, if $\Xi = \{?p, ?q\}$, any of the following is a Ξ-answer set: $\{p, q\}, \{\neg p, q\}, \{p, \neg q\}, \{\neg p, \neg q\}$.

By a *binary question* we mean a question which has exactly two direct answers. We say that entailment in $\mathcal{L}$, $\models_\mathcal{L}$, is *compact* iff $X \models_\mathcal{L} A$ yields $X^* \models_\mathcal{L} A$ for some finite subset X^* of X. By a *valid* d-wff we mean a d-wff which is true in each admissible partition of the language.

Theorem 7.25. *Let $\mathcal{L}$ be a language of the considered kind such that:*
(a) *entailment in $\mathcal{L}$ is compact,*
(b) *for each d-wff A of $\mathcal{L}$ there exists a d-wff, $\overline{A}$, of the language such that $\overline{A}$ eliminates A in $\mathcal{L}$ and $\emptyset \models_\mathcal{L} \{A, \overline{A}\}$; moreover, $\{A, \overline{A}\}$ is the set of direct answers to a question of $\mathcal{L}$ given that A is a direct answer to a question of $\mathcal{L}$,*
(c) *at least one d-wff of $\mathcal{L}$ is valid.*

If $\mathbf{E}_\mathcal{L}(X, Q)$, then there exists a non-empty finite set, Ξ, of binary questions of $\mathcal{L}$ that fulfils the following conditions:

1. *each question of Ξ is evoked in $\mathcal{L}$ by X,*
2. *each question of Ξ is strongly implied in $\mathcal{L}$ by Q on the basis of X, and*
3. *for each Ξ-answer set Z: $X \cup Z$ entails in $\mathcal{L}$ a direct answer to Q.*

Proof. For conciseness, we will be omitting references to $\mathcal{L}$.

Let Λ be the family of all "inconsistent" sets of d-wffs of the language. More formally:
$$\Lambda = \{X : X \not\subset \mathsf{T}_\mathsf{P} \text{ for each admissible partition } \mathsf{P}\}$$
If entailment is compact and the assumptions (b) and (c) hold, then the following is true:

($\star$) $X \in \Lambda$ iff for some non-empty finite subset X^* of X: $X^* \in \Lambda$

For, if $X \in \Lambda$, then $X \models C$, where C is a formula eliminated by a valid d-wff (and hence $\{C\} \in \Lambda$). By compactness of $\models$ we get $X^* \models C$ for some finite subset X^* of X. Clearly, $X^* \in \Lambda$ and $X^* \neq \emptyset$.

It follows that mc-entailment is compact. To see this, assume that $X \mid\models Y$. Let $\overline{Y} = \{\overline{B} : B \in Y\}$. Obviously, $X \mid\models Y$ iff $(X \cup \overline{Y}) \in \Lambda$. Therefore, by ($\star$), $Z \in \Lambda$ for some non-empty finite subset Z of $X \cup \overline{Y}$. But $Z = X_1 \cup \overline{Y_1}$, where $X_1 \subset X$, $\overline{Y_1} \subset \overline{Y}$, and at least one of $X_1, \overline{Y_1}$ is non-empty. Hence $X_1 \mid\models Y_1$, where X_1, Y_1 are finite subsets of X and Y, respectively. Thus $\mid\models$ is compact.

Let $\mathbf{E}(X, Q)$. Then $X \mid\models \mathsf{d}Q$ and hence, by compactness of mc-entailment, there exists a non-empty family ϱ of finite subsets of $\mathsf{d}Q$ such that $X \mid\models Z$ for any $Z \in \varrho$. Clearly, ϱ contains minimal elements, that is, finite subsets of $\mathsf{d}Q$ which do not have proper subsets mc-entailed by X. On the other hand, each set in ϱ has at least two elements; otherwise $\mathbf{E}(X, Q)$ would not hold.

Let $Z^* = \{A_1, A_2, \ldots, A_n\}$ be an arbitrary but fixed minimal element of ϱ. Clearly, $n > 1$. For each $A_i \in Z^*$ we fix the corresponding d-wff $\overline{A_i}$. Then we consider the following family of sets of d-wffs:

$$\{\{A_1, \overline{A_1}\}, \ldots, \{A_{n-1}, \overline{A_{n-1}}\}\} \tag{7.61}$$

For each element $\{A_i, \overline{A_i}\}$ of (7.61) we fix a question, Q_i, such that $dQ_i = \{A_i, \overline{A_i}\}$. We designate the corresponding set of questions by Ξ.

Let $1 \leq i \leq n-1$. Since $\emptyset \Vdash \{A_i, \overline{A_i}\}$, we get $X \Vdash dQ_i$. If $\mathbf{E}(X, Q)$, then $X \not\models A_i$. Suppose that $X \models \overline{A_i}$. Since $X \Vdash Z^*$ and $\overline{A_i}$ eliminates A_i, we get $X \Vdash Z^* \setminus \{A_i\}$. Hence Z^* is not a minimal element of ϱ. A contradiction. Thus we may conclude that each question of Ξ is evoked by X.

Since $\emptyset \Vdash \{A_i, \overline{A_i}\}$, the first clause of Definition 7.18 of strong erotetic implication is fulfilled by any question in Ξ. As for the second clause, observe, first, that $X \cup \{A_i\} \models A_i$, but, due to the fact that Q is evoked by X, also $X \not\models A_i$. Now consider $\overline{A_i}$. Since $X \Vdash Z^*$, where $Z^* \subset dQ$, and $\overline{A_i}$ eliminates A_i, we get $X \cup \{\overline{A_i}\} \Vdash Z^* \setminus \{A_i\}$. As Z^* is a minimal element of ϱ, we also have $X \not\Vdash Z^* \setminus \{A_i\}$. Therefore each question of Ξ is strongly implied, on the basis of X, by the initial question Q.

Let ξ be a Ξ-answer set. There are two options: (a) $A_i \in \xi$ for some $1 \leq i \leq n-1$, or (b) $\xi = \{\overline{A_1}, \ldots, \overline{A_{n-1}}\}$. If (a) holds, then $X \cup \xi$ entails A_i, which is also a direct answer to Q. Assume that (b) holds. We have $X \Vdash \{A_1, \ldots, A_n\}$. Thus $X \cup \xi$ entails A_n, which, again, is a direct answer to Q. □

Note that the compactness assumption is dispensable if the set of direct answers to Q is finite. The same holds true for the assumption (c). Needless to say, the first part of the assumption (b) is fulfilled if the language contains negation classically construed.

A philosophical comment is this: when the cognitive goal is set by a conclusion of a valid erotetic inference of the first kind, it can[8] be accomplished by answering question(s) arrived at by means of valid erotetic inference(s) of the second kind. However, it cannot be said that this is always the best way of reaching the goal. We will come back to this issue in Chapter 8.

Remark. Theorem 7.25 can be rephrased by using the concept of (generalized) reducibility of questions to sets of questions. A question Q is reducible to a set of questions Ξ on the basis of a set of d-wffs X iff (a) for each admissible partition P of the language considered: if Q is sound in P and the d-wffs of X are true in P, then each question of Ξ is sound in P, (b) for each Ξ-answer set ξ: $X \cup \xi$ entails some direct answer to Q which is not entailed by X alone, and (c) no question in Ξ has more direct answers than Q has. Thus the claim of Theorem 7.25 is: *given that conditions (a), (b), and (c) are met, a question Q evoked by a set of d-wffs X is reducible, on the basis of X, to a finite set of binary questions which are both evoked by X and strongly implied by Q along*

---

[8] Given that the assumptions of Theorem 7.25 are met or the evoked question has a finite set of direct answers.

with X. For an overview of various reducibility results pertaining to first-order languages augmented with questions see Leśniewski and Wiśniewski (2001).[9]

---

[9] Besides generalized reducibility, the paper Leśniewski and Wiśniewski (2001) considers the case of *plain* reducibility, that is, reducibility not relativized to sets of d-wffs. Both concepts were introduced in Wiśniewski (1994b). Generalized reducibility is extensively studied in Leśniewski (1997); see also Leśniewski (2000).

8

Socratic Transformations

Is it possible to solve a problem by performing valid erotetic inferences only? At first sight this question is rhetorical since the only plausible answer seems to be "No". In order to solve a problem one has to extract information and, although questions are good tools for doing it, it is the answers that count. However, there is an old idea, going back at least to Socrates, according to which questions can be resolved by transforming them into questions whose answers are, in a sense, evident. In this chapter we show how the idea can be explicated in terms of IEL.

8.1 Language $\mathcal{L}^?_{\vdash cpl}$ again

Let us come back to the language $\mathcal{L}^?_{\vdash opl}$, whose syntax was presented in section 2.4.4 of Chapter 2, and semantics in section 3.1.3 of Chapter 3. In order to make this chapter self-contained we recall the basic notions and we introduce some notational conventions.

8.1.1 Syntax

Atomic d-wffs of $\mathcal{L}^?_{\vdash cpl}$ are CPL-sequents of the form $S \vdash A$, where S is a finite sequence of CPL-formulas, and A is a single CPL-formula. Compound d-wffs of $\mathcal{L}^?_{\vdash cpl}$ are built from the atomic d-wffs by means of & and/or ng. As for CPC-formulas, we distinguish α and β formulas, and make some assignments displayed in Table 8.1.

Table 8.1. α/β formulas.

α	α_1	α_2	β	β_1	β_2	β_1^*
$A \wedge B$	A	B	$\neg(A \wedge B)$	$\neg A$	$\neg B$	A
$\neg(A \vee B)$	$\neg A$	$\neg B$	$A \vee B$	A	B	$\neg A$
$\neg(A \to B)$	A	$\neg B$	$A \to B$	$\neg A$	B	A

Terminology: sequents. Till the end of this chapter, unless otherwise stated, by sequents we mean atomic d-wffs of $\mathcal{L}^?_{\vdash cpl}$.

Questions of $\mathcal{L}^?_{\vdash cpl}$ are of the form $?(\Phi)$, where Φ is a non-empty finite sequence of *atomic* d-wffs of $\mathcal{L}^?_{\vdash cpl}$, that is, sequents. Direct answers to a question:

$$?(S_1 \vdash A_1, \ldots, S_n \vdash A_n) \tag{8.1}$$

are of the forms (we omit unnecessary parentheses; ng stands for negation in $\mathcal{L}^?_{\vdash cpl}$):

$$S_1 \vdash A_1 \& \ldots \& S_n \vdash A_n \tag{8.2}$$

$$ng(S_1 \vdash A_1 \& \ldots \& S_n \vdash A_n) \tag{8.3}$$

Intuitively, the turnstile $\vdash$ stands for CPL entailment/derivability.

Notation: concatenations of sequences of sequents. We use the semicolon as the concatenation-sign for sequences of sequents. Hence

$$\Phi \; ; \; \phi$$

denotes the concatenation of a sequence of sequents Φ and the one-term sequence $\langle \phi \rangle$, where ϕ is a sequent.[1] The expression:

$$\Phi \; ; \; \phi \; ; \; \Psi$$

refers to the concatenation of $\Phi \; ; \; \phi$ and a sequence of sequents Ψ. Any of Φ, Ψ may be empty.

Observe that $\phi_1, \ldots, \phi_n$ can be displayed as:

$$\phi_1 \; ; \; \ldots \; ; \; \phi_n$$

Thus when $\Phi = \langle S_1 \vdash A_1 \rangle, \ldots, \langle S_n \vdash A_n \rangle$, the corresponding question can be written as follows:

$$?(S_1 \vdash A_1 \; ; \; \ldots \; ; \; S_n \vdash A_n) \tag{8.4}$$

and we will proceed that way.[2] The sequents in (8.4) are called *constituents* of the question.

If $\Phi = \langle S \vdash A \rangle$, that is, Φ is the one-term sequence $\langle S \vdash A \rangle$, we write the question as:

$$?(S \vdash A) \tag{8.5}$$

and we say that the question is based on a (single) sequent.

8.1.2 Semantics

Admissible partitions of $\mathcal{L}^?_{\vdash cpl}$ are defined as follows (we repeat Definition 3.6):

Definition 8.1 (*Admissible partitions of $\mathcal{L}^?_{\vdash cpl}$*). *A partition* $\mathsf{P} = \langle \mathsf{T_P}, \mathsf{U_P} \rangle$ *of $\mathcal{L}^?_{\vdash cpl}$ is admissible iff the following conditions hold:*[3]

---

[1] We omit, here and below, the signs $\langle, \rangle$ when referring to one-term sequences.
[2] Obeying this convention seems to facilitate reading; compare $A, B \vdash C \; ; \; D \vdash A$ with $A, B \vdash C, D \vdash A$.
[3] As before (cf. page 28), ′ is the concatenation-sign for sequences of CPL-formulas; $\mathfrak{r}, \mathfrak{u}$ are metalanguage variables for d-wffs of $\mathcal{L}^?_{\vdash cpl}$.

1. $\ulcorner S \vdash \alpha \urcorner \in \mathsf{T_P}$ iff $\ulcorner S \vdash \alpha_1 \urcorner \in \mathsf{T_P}$ and $\ulcorner S \vdash \alpha_2 \urcorner \in \mathsf{T_P}$;
2. $\ulcorner S' T \vdash \beta \urcorner \in \mathsf{T_P}$ iff $\ulcorner S' \beta_1^* ' T \vdash \beta_2 \urcorner \in \mathsf{T_P}$;
3. $\ulcorner S' \alpha ' T \vdash C \urcorner \in \mathsf{T_P}$ iff $\ulcorner S' \alpha_1 ' \alpha_2 ' T \vdash C \urcorner \in \mathsf{T_P}$;
4. $\ulcorner S' \beta ' T \vdash C \urcorner \in \mathsf{T_P}$ iff $\ulcorner S' \beta_1 ' T \vdash C \urcorner \in \mathsf{T_P}$ and $\ulcorner S' \beta_2 ' T \vdash C \urcorner \in \mathsf{T_P}$;
5. $\ulcorner S \vdash \neg\neg A \urcorner \in \mathsf{T_P}$ iff $\ulcorner S \vdash A \urcorner \in \mathsf{T_P}$;
6. $\ulcorner S' \neg\neg A ' T \vdash B \urcorner \in \mathsf{T_P}$ iff $\ulcorner S' A ' T \vdash B \urcorner \in \mathsf{T_P}$;
7. $\ulcorner \mathfrak{r} \,\&\, \mathfrak{u} \urcorner \in \mathsf{T_P}$ iff $\mathfrak{r} \in \mathsf{T_P}$ and $\mathfrak{u} \in \mathsf{T_P}$;
8. if $\mathfrak{u} \notin \mathsf{T_P}$, then $\ulcorner ng\, \mathfrak{u} \urcorner \in \mathsf{T_P}$;
9. if $\mathfrak{u} \in \mathsf{T_P}$, then $\ulcorner ng\, \mathfrak{u} \urcorner \notin \mathsf{T_P}$.

A d-wff $\mathfrak{u}$ is entailed in $\mathcal{L}^?_{\vdash cpl}$ by a d-wff $\mathfrak{r}$ iff there is no admissible partition $\mathsf{P} = \langle \mathsf{T_P}, \mathsf{U_P} \rangle$ of $\mathcal{L}^?_{\vdash cpl}$ such that $\mathfrak{r} \in \mathsf{T_P}$ and $\mathfrak{u} \in \mathsf{U_P}$.

A sequent $S \vdash A$ is said to be CPL-*valid* iff A is true in each admissible partition of $\mathcal{L}_{cpl}$ in which all the terms of S are true. In other words, $S \vdash A$ is CPL-valid just in case each CPL-valuation that makes true all the terms of S makes A true as well. Thus CPL-validity of $S \vdash A$ amounts to the CPL-entailment/derivability of A from the set of terms of S.

8.2 From questions to questions

8.2.1 Some examples

Let us start with examples.

Example 8.2. We have:

$$\mathbf{Im}_{\mathcal{L}^?_{\vdash cpl}}(?(p \to q \vdash \neg q \to \neg p), ?(p \to q, \neg q \vdash \neg p)) \qquad (8.6)$$

(8.6) holds because affirmative answers to:

$$?(p \to q \vdash \neg q \to \neg p) \qquad (8.7)$$

$$?(p \to q, \neg q \vdash \neg p) \qquad (8.8)$$

entail each other in $\mathcal{L}^?_{\vdash cpl}$, and similarly for negative answers. To see this, let us assume, first, that the affirmative answer to (8.7), that is:

$$p \to q \vdash \neg q \to \neg p \qquad (8.9)$$

belongs to $\mathsf{T_P}$ for an arbitrary but fixed admissible partition P of $\mathcal{L}^?_{\vdash cpl}$. Then, by clause (2) of Definition 8.1, the d-wff:

$$p \to q, \neg q \vdash \neg p \qquad (8.10)$$

is in $\mathsf{T_P}$ as well. But (8.10) is the affirmative answer to question (8.8).

Now assume that (8.10) belongs to $\mathsf{T_P}$, again for an arbitrary but fixed admissible partition P. Thus, by the clause (2), the affirmative answer to question (8.7), that is, the d-wff (8.9), belongs to $\mathsf{T_P}$ as well.

The reasoning concerning negative answers goes along similar lines (we, additionally, use clauses (8) and (9)).

Observe that (8.6) is a special case of:

$$\mathbf{Im}_{\mathcal{L}^?_{\vdash cpl}}(?(\Phi\,;\,S \vdash \beta;\,\Psi), ?(\Phi\,;\,S\,'\,\beta_1^* \vdash \beta_2\,;\,\Psi)) \qquad (8.11)$$

For, $\neg q \to \neg p$ is a β-formula, $\beta_1^* = \neg q$, and $\beta_2 = \neg p$. The reasoning that proves (8.11) is similar to that justifying (8.6); the implying question and the implied question differ with respect to one constituent only.

Since erotetic implication determines validity of erotetic inferences of the second kind, we have:

Corollary 8.3. *An inference from a question of the form:*

$$?(\Phi\,;\,S \vdash \beta\,;\,\Psi) \qquad (8.12)$$

to the corresponding question of the form:

$$?(\Phi\,;\,S\,'\,\beta_1^* \vdash \beta_2\,;\,\Psi) \qquad (8.13)$$

is valid.

Thus, in particular, the transition from question (8.7) to question (8.8) is a valid erotetic inference.

Validity of erotetic inferences is one thing, CPL-validity of sequents being constituents of questions is another.

One can prove:

Corollary 8.4. *Each constituent of a question of the form (8.12) is a CPL-valid sequent if, and only if each constituent of the corresponding question of the form (8.13) is a CPL-valid sequent.*

The idea of the proof is simple: (8.12) and (8.13) differ with respect to only one constituent/sequent. But, due to clause (2) of Definition 8.1, $\ulcorner S \vdash \beta \urcorner \in \mathsf{T_P}$ iff $\ulcorner S\,'\,\beta_1^* \vdash \beta_2 \urcorner \in \mathsf{T_P}$, for any admissible partition P.

Generally speaking, the transition from a question of the form (8.12) to the corresponding question of the form (8.13) is a valid erotetic inference in which, at a deeper level, *joint* CPL-validity of sequents is preserved *in both directions*.

Example 8.5. The following holds:

$$\mathbf{Im}_{\mathcal{L}^?_{\vdash cpl}}(?(p \to q, \neg q \vdash \neg p)\,,\,?(\neg p, \neg q \vdash \neg p\,;\,q, \neg q \vdash \neg p)) \qquad (8.14)$$

Let $\mathsf{T_P}$ be an admissible partition of $\mathcal{L}^?_{\vdash cpl}$. Assume that (8.10), being the affirmative answer to question (8.8):

$$?(p \to q, \neg q \vdash \neg p)$$

is in $\mathsf{T_P}$. Hence, by clause (4) of Definition 8.1, the following:

$$\neg p, \neg q \vdash \neg p \qquad (8.15)$$

are in T_P as well. Thus, by the clause (7), the d-wff:

$$q, \neg q \vdash \neg p \tag{8.16}$$

$$(\neg p, \neg q \vdash \neg p) \,\&\, (q, \neg q \vdash \neg p) \tag{8.17}$$

belongs to T_P. But (8.17) is the affirmative answer to the question:

$$?(\neg p, \neg q \vdash \neg p \,;\, q, \neg q \vdash \neg p) \tag{8.18}$$

Now assume that (8.17) is in T_P. Therefore, by the clause (7), the d-wffs (8.15) and (8.16) are in T_P. Hence, due to the clause (4), (8.10), the affirmative answer to question (8.8), belongs to T_P.

Thus affirmative answers to questions (8.8) and (8.18) entail each other. It is easily seen that the same holds for negative answers. Therefore (8.8) implies (8.18) in $\mathcal{L}^?_{\vdash cpl}$.

Observe that (8.14) is a special case of:

$$\mathbf{Im}_{\mathcal{L}^?_{\vdash cpl}}(?(\Phi \,;\, S\,'\, \beta\,'\, T \vdash B;\, \Psi), ?(\Phi \,;\, S\,'\, \beta_1\,'\, T \vdash B \,;\, S\,'\, \beta_2\,'\, T \vdash B \,;\, \Psi)) \tag{8.19}$$

which is also true. Hence:

Corollary 8.6. *An inference from a question of the form:*

$$?(\Phi \,;\, S\,'\, \beta\,'\, T \vdash B \,;\, \Psi) \tag{8.20}$$

to the corresponding question of the form:

$$?(\Phi \,;\, S\,'\, \beta_1\,'\, T \vdash B \,;\, S\,'\, \beta_2\,'\, T \vdash B \,;\, \Psi) \tag{8.21}$$

is valid.

At the same time we have:

Corollary 8.7. *Each constituent of a question of the form (8.20) is a CPL-valid sequent if, and only if each constituent of the corresponding question of the form (8.21) is a CPL-valid sequent.*

The general comment to be made is analogous as before.

Let us designate by s_1 the following sequence of questions:

1. $?(p \to q \vdash \neg q \to \neg p)$
2. $?(p \to q, \neg q \vdash \neg p)$
3. $?(\neg p, \neg q \vdash \neg p \,;\, q, \neg q \vdash \neg p)$

Each consecutive term of the sequence is erotetically implied by the term that immediately precedes it. On the other hand, due to corollaries 8.4 and 8.7, a consecutive term comprises CPL-valid sequent(s) if, and only if its immediate predecessor comprises such sequent(s).

Now let us consider a similar sequence of questions, s_2:

1. $?(p \to q \vdash \neg p \to \neg q)$

2. $?(p \rightarrow q, \neg p \vdash \neg q)$
3. $?(\neg p, \neg p \vdash \neg q \; ; \; q, \neg p \vdash \neg q)$

Everything what has been said above about s_1 can be repeated concerning s_2. However, unlike s_2, the sequence s_1 ends with a question which is, in a sense, rhetorical: no one doubts that a CPL-sequent which has the same formula on both sides of the turnstile is valid, and that a CPL-sequent which has a formula and its negation left to the turnstile is valid. The last question of sequence s_1 reads:

Is it the case that: $\neg p$ is CPL-entailed by $\neg p, \neg q$ and
$\neg p$ is CPL-entailed by $q, \neg q$?

and its affirmative answer is evident. On the other hand, due to the transmission of joint CPL-validity of sequents from bottom to top, it follows without the need of any further reasoning that the sequent occurring in the first question is CPL-valid, or, to put it differently, that the answer to the first question must be affirmative. In other words, the initial issue is solved by pure questioning: the only operations needed are performed upon questions which, let us stress, need not be answered.

The case of sequence s_2 is similar, though the solution provided is negative. The last question of the sequence reads:

Is it the case that: $\neg q$ is CPL-entailed by $\neg p, \neg p$ and
$\neg q$ is CPL-entailed by $q, \neg p$?

and it is evident that the answer is negative, which, due to the transmission of joint CPL-validity from bottom to top, amounts to the negative answer to the first question.

Both sequences, s_1 and s_2, are *Socratic transformations*, but the first sequence, s_1, is a successful transformation, that is, a *Socratic proof*.

8.2.2 $\mathbb{E}^*$: An erotetic calculus for CPL

Statements (8.11) and (8.19) specified above are examples of metatheorems stating what questions are implied by what questions. Metatheorems of this kind enable us to formulate *erotetic rules*, being rules of transitions from questions to questions. A set of rules of this kind together with a characterization of *basic sequents* (cf. below) constitute an *erotetic calculus*.

Here are the rules of the erotetic calculus $\mathbb{E}^*$ for CPL.[4] The letters S, T, U, W stand for finite (possibly empty) sequences of CPL-formulas and ' is the concatenation-sign for these sequences. As before, the letters Φ, Ψ are metalanguage variables for finite (again, possibly empty) sequences of CPL-sequents, and the semicolon is used as the concatenation-sign for sequences of CPL-sequents. One-term sequences are represented by their terms.

---

[4] Proposed in Wiśniewski (2004b).

$$\mathbf{L}_\alpha : \frac{?(\varPhi\,;\,S'\,\alpha'\,T \vdash C\,;\,\varPsi)}{?(\varPhi\,;\,S'\,\alpha_1'\,\alpha_2'\,T \vdash C\,;\,\varPsi)} \qquad \mathbf{R}_\alpha : \frac{?(\varPhi\,;\,S \vdash \alpha\,;\,\varPsi)}{?(\varPhi\,;\,S \vdash \alpha_1\,;\,S \vdash \alpha_2\,;\,\varPsi)}$$

$$\mathbf{L}_\beta : \frac{?(\varPhi\,;\,S'\,\beta'\,T \vdash C\,;\,\varPsi)}{?(\varPhi\,;\,S'\,\beta_1'\,T \vdash C\,;\,S'\,\beta_2'\,T \vdash C\,;\,\varPsi)} \qquad \mathbf{R}_\beta : \frac{?(\varPhi\,;\,S \vdash \beta\,;\,\varPsi)}{?(\varPhi\,;\,S'\,\beta_1^* \vdash \beta_2\,;\,\varPsi)}$$

$$\mathbf{L}_{\neg\neg} : \frac{?(\varPhi\,;\,S'\,\neg\neg A'\,T \vdash C\,;\,\varPsi)}{?(\varPhi\,;\,S'\,A'\,T \vdash C\,;\,\varPsi)} \qquad \mathbf{R}_{\neg\neg} : \frac{?(\varPhi\,;\,S \vdash \neg\neg A\,;\,\varPsi)}{?(\varPhi\,;\,S \vdash A\,;\,\varPsi)}$$

The letters "**L**" and "**R**" indicate that the appropriate rule "operates" on the left or right side of the turnstile $\vdash$. The second part of the rule's name indicates the form of a CPL-formula acted upon. For instance, rule $\mathbf{R}_\alpha$ operates on an α-formula occurring on the right side of the turnstile.

$\mathbf{R}_\alpha$ and $\mathbf{L}_\beta$ are *branching rules*, as the resulting "question-conclusion" has more constituents than the "question-premise". The remaining erotetic rules: $\mathbf{L}_\alpha$, $\mathbf{R}_\beta$, $\mathbf{L}_{\neg\neg}$ and $\mathbf{R}_{\neg\neg}$ are non-branching.

The rules of $\mathbb{E}^*$ are designed in such a way that each constituent of the "question-conclusion" is a CPC-valid sequent if and only if each constituent of the "question-premise" is a CPC-valid sequent. In other words, the transmission of joint CPL-validity of sequents holds in both directions. On the other hand, each application of a rule of $\mathbb{E}^*$ retains validity of the corresponding erotetic inference. Corollaries 8.3, 8.4, 8.6 and 8.7 justify the above claims for rules $\mathbf{R}_\beta$ and $\mathbf{L}_\beta$. As for the remaining rules, the corollaries needed can be proven in a similar way, by using Definition 8.1. For details see Wiśniewski (2004b).

The concept of Socratic transformation is characterized by:

Definition 8.8 (*Socratic transformation*). *A sequence $\langle s_1, s_2, \ldots \rangle$ of questions is a Socratic transformation of a question $?(\varPhi)$ via the rules of $\mathbb{E}^*$ iff the following conditions hold:*

1. $s_1 = ?(\varPhi)$,
2. s_i, *where $i > 1$, results from s_{i-1} by an application of a rule of $\mathbb{E}^*$.*

The sequences s_1 and s_2 considered above (see page 93) are examples of Socratic transformations of questions based on single sequents.

Since an application of a rule of $\mathbb{E}^*$ is a transition from a question to a question (erotetically) implied by the previous one, we get:

Corollary 8.9. *Each step of a Socratic transformation via the rules of $\mathbb{E}^*$ is a valid erotetic inference.*

Let us now introduce:

Definition 8.10 (*Basic sequents*). *A sequent ϕ is basic iff ϕ is of one of the following forms:*

1. $T'\,B'\,U \vdash B$, *or*
2. $T'\,B'\,U'\,\neg B'\,W \vdash C$, *or*
3. $T'\,\neg B'\,U'\,B'\,W \vdash C$.

Note that any of T, U, W can be empty.

Thus a basic sequent either has its consequent among the premises, or has a CPL-formula and its negation among the premises. Given this, the following comes with no surprise:

Corollary 8.11. *Each basic sequent is CPL-valid.*

Now we are ready to introduce the concept of Socratic proof.

Definition 8.12 (*Socratic proof*). *A Socratic proof in $\mathbb{E}^*$ of a sequent $S \vdash A$ is a finite Socratic transformation of the question $?(S \vdash A)$ via the rules of $\mathbb{E}^*$ such that each constituent of the last question of the transformation is a basic sequent.*

The sequence s_1 above is a Socratic proof in $\mathbb{E}^*$, while the sequence s_2 is not. Here are further examples of Socratic proofs. For transparency, we highlight the constituents which a given rule acts upon and we put the name of the rule to the right.

Example 8.13.

$\vdash ?(\ (p \to q) \wedge (q \to r) \to (p \to r) \ )$ $\quad\quad\quad\quad\quad\quad\quad\quad\quad$ **R**$_\beta$
$?(\ (p \to q) \wedge (q \to r) \ \vdash \ p \to r)$ $\quad\quad\quad\quad\quad\quad\quad\quad\quad\quad\quad\quad$ **L**$_\alpha$
$?(p \to q, q \to r \ \vdash \ p \to r \ )$ $\quad\quad\quad\quad\quad\quad\quad\quad\quad\quad\quad\quad\quad\quad\quad$ **R**$_\beta$
$?(\ p \to q \ , \ q \to r, p \ \vdash \ r)$ $\quad\quad\quad\quad\quad\quad\quad\quad\quad\quad\quad\quad\quad\quad\quad$ **L**$_\beta$
$?(\neg p, q \to r, p \ \vdash \ r \ ; \ q \ , \ q \to r \ , \ p \ \vdash \ r)$ $\quad\quad\quad\quad\quad\quad\quad\quad$ **L**$_\beta$
$?(\neg p, q \to r, p \ \vdash \ r \ ; \ q, \neg q, p \ \vdash \ r \ ; \ q, r, p \ \vdash \ r)$

Example 8.14.

$?(\ \neg(p \vee q) \ \vdash \ \neg p \wedge \neg q)$ $\quad\quad\quad\quad\quad\quad\quad\quad\quad\quad\quad\quad\quad\quad$ **L**$_\alpha$
$?(\neg p, \neg q \ \vdash \ \neg p \wedge \neg q \ )$ $\quad\quad\quad\quad\quad\quad\quad\quad\quad\quad\quad\quad\quad\quad\quad$ **R**$_\alpha$
$?(\neg p, \neg q \ \vdash \ \neg p \ ; \ \neg p, \neg q \ \vdash \ \neg q)$

The following holds:

Lemma 8.15. *If sequent $S \vdash A$ has a Socratic proof in $\mathbb{E}^*$, then $S \vdash A$ is CPL-valid.*

The reason is simple: basic sequents are CPL-valid, and the rules of $\mathbb{E}^*$ preserve joint CPL-validity of sequents from bottom to top.

Remark. Thus when a Socratic proof is found, there is no need for any further deductive moves aimed at justifying validity of the sequent just (Socratically) proven. Note also that the argument for validity has the form of a sequence of *questions* and that a reasoning towards proof-search operates on questions only. Moreover, one does not have to answer intermediate questions to arrive at a proof, while the last question of a Socratic proof is, in a sense, rhetorical.

The following is true as well:

Lemma 8.16. *If sequent $S \vdash A$ is CPL-valid, then $S \vdash A$ has a Socratic proof in $\mathbb{E}^*$.*

Generally speaking, Lemma 8.16 holds because the rules of $\mathbb{E}^*$ are eliminative, that is, each application of a rule eliminates either an occurrence of a binary connective or two occurrences of negation. Moreover, there is no rule that acts upon negated propositional variables. As a result, each Socratic transformation in $\mathbb{E}^*$ is finite and either ends with a question which involves only literals (i.e. propositional variables and/or their negations) in the constituents of the last question, or can be extended to such transformation. Now suppose that $S \vdash A$ is CPL-valid, but no Socratic transformation of the sequent via the rules of $\mathbb{E}^*$ is a Socratic proof. Consider an arbitrary but fixed Socratic transformation of $S \vdash A$ that ends with a question, $Q^{\#}$, which involves only literals. Since the transformation is not a Socratic proof, at least one constituent of $Q^{\#}$ is not a basic sequent. But, as the CPL-formulas occurring in the sequent/constituent are literals, this yields that there is no literal that occurs on both sides of the turnstile, and there is no propositional variable which occurs together with its negation left to the turnstile. Hence, for obvious reasons, the sequent is not CPL-valid. Now recall that the rules of $\mathbb{E}^*$ (also) preserve joint validity of sequents from top to bottom. Since it is not the case that *all* sequents/constituents of $Q^{\#}$ are CPL-valid, the same holds true for any other question of the transformation, the first question included. The first question, $?(S \vdash A)$, is based on a single sequent. Therefore the sequent $S \vdash A$ is not CPL-valid. We arrive at a contradiction. Hence $S \vdash A$ has a Socratic proof in $\mathbb{E}^*$.

By Lemma 8.15 and Lemma 8.16 we get:

Theorem 8.17. *A sequent is CPL-valid iff the sequent has a Socratic proof in $\mathbb{E}^*$.*

For a detailed proof see Wiśniewski (2004b).

Recall that the sequents considered above are single-conclusioned: they have single CPL-formulas right to the turnstile and a finite (possibly empty) sequence of CPL-formulas left to the turnstile. Let us also stress that sequents and questions are expressions of the *object-level* language $\mathcal{L}_{\vdash cpl}$.

8.2.3 Other erotetic calculi

$\mathbb{E}^*$ is only one of the erotetic calculi developed so far. There exist erotetic calculi for First-order Logic (cf. Wiśniewski and Shangin (2006)), some paraconsistent propositional logics (cf. Wiśniewski et al. (2005)), and for a wide class of modal propositional logics (cf. Leszczyńska (2004), Leszczyńska (2007), Leszczyńska-Jasion (2008), Leszczyńska-Jasion (2009)). In each case completeness theorems are proven; in some cases decision procedures are described. However, erotetic calculi for logics other than CPL require a more sophisticated languages. Generally speaking, their erotetic parts resemble that of $\mathcal{L}^?_{\vdash cpl}$, but the declarative parts are more complex.

8.2.4 Erotetic calculi vs. sequent calculi

Socratic proofs can be transformed into Gentzen-style proofs in "parallel" sequent calculi. For example, the sequent calculus $\mathbb{G}^*$ parallel to the erotetic calculus $\mathbb{E}^*$ has the following rules:

$$\mathbb{G}^*\mathbb{L}_\alpha : \frac{S\,'\,\alpha_1\,'\,\alpha_2\,'\,T \vdash C}{S\,'\,\alpha\,'\,T \vdash C} \qquad \mathbb{G}^*\mathbb{R}_\alpha : \frac{S \vdash \alpha_1 \quad S \vdash \alpha_2}{S \vdash \alpha}$$

$$\mathbb{G}^*\mathbb{L}_\beta : \frac{S\,'\,\beta_1\,'\,T \vdash C \quad S\,'\,\beta_2\,'\,T \vdash C}{S\,'\,\beta\,'\,T \vdash C} \qquad \mathbb{G}^*\mathbb{R}_\beta : \frac{S\,'\,\beta_1^* \vdash \beta_2}{S \vdash \beta}$$

$$\mathbb{G}^*\mathbb{L}_{\neg\neg} : \frac{S\,'\,A\,'\,T \vdash B}{S\,'\,\neg\neg A\,'\,T \vdash B} \qquad \mathbb{G}^*\mathbb{R}_{\neg\neg} : \frac{S \vdash A}{S \vdash \neg\neg A}$$

Axioms of the calculus are the basic sequents specified in Definition 8.10. Calculus $\mathbb{G}^*$ provides a sound and complete formalization of CPL.[5] Again, the sequents operated with have *sequences* of formulas left to the turnstile and *single* formulas right to the turnstile. On the other hand, $\mathbb{G}^*$ has no primary structural rules. A translation of a Socratic proof in $\mathbb{E}^*$ into a proof in $\mathbb{G}^*$ can be made algorithmically; the same pertains to the erotetic calculus for FoL and its parallel sequent calculus. For details see Leszczyńska-Jasion et al. (2013).

Finally, let us clarify the following. One can suspect that an erotetic calculus is, in essence, nothing more than a sequent calculus presented upside-down and ornamented with question marks. This is clearly wrong. For example, take the usual rule for the introduction of a disjunction and its reverse:

$$\frac{S \vdash A}{S \vdash A \vee B} \qquad \frac{S \vdash A \vee B}{S \vdash A}$$

One cannot say that an inference from $?(S \vdash A)$ to $?(S \vdash A \vee B)$ is valid, and similarly for an inference from $?(S \vdash A \vee B)$ to $?(S \vdash A)$.[6] More importantly, CPL-validity of $S \vdash A \vee B$ does not warrant CPL-validity of $S \vdash A$. For instance, $p \vdash q \vee p$ is valid, but $p \vdash q$ is not.

There are close affinities between erotetic calculi and sequent calculi with semantically reversible rules.[7] Both enable a modelling of "equivalence reasoning" in proof-search. The former, however, conceives proof-search as a Socratic transformation performed in an object-level language. Yet, one cannot take *any* sequent calculus and then, by syntactic tricks, turn it into an erotetic calculus.

8.2.5 Internal question processing

The received view on question asking and answering favours a dyadic account. It is assumed that there are two parties, a questioner and an answerer. The

---
[5] Cf. Wiśniewski (2004b) or Leszczyńska-Jasion et al. (2013).
[6] Because $ng(S \vdash A \vee B)$ entails (in $\mathcal{L}^?_{\vdash cpl}$) neither $S \vdash A$ nor $ng(S \vdash A)$. Similarly, $ng(S \vdash A)$ entails neither $S \vdash A \vee B$ nor $ng(S \vdash A \vee B)$.
[7] For the latter see Rasiowa and Sikorski (1963), and Negri and von Plato (2001).

former asks a question, whereas the role of the latter is to provide an answer to the question. The parties are defined by their roles and, in some cases, both roles can be played, consecutively, by the same agent; the phenomenon of "asking questions to myself" is a case in point here. Moreover, an answerer need not be a human: even eliciting information from Nature is sometimes modelled in the dyadic perspective.[8]

However, the above account neglects the phenomenon of *internal question processing*. When an agent is supposed to answer a question or solve a problem, but he/she cannot accomplish the task by means of informational resources which are directly accessible to him/her, it often happens that he/she internally processes the initial question. The outcome is either a new question concerning the subject matter or a preliminary strategy of reducing the initial question into auxiliary questions. In both cases erotetic inferences play a substantial role. When answers to questions raised are still inaccessible, the process goes further in an analogous way, possibly with the help of data just collected.

Internal question processing (hereafter: IQP) can be *ultimate* or *distributed*.

It happens that one arrives at a satisfactory answer without sending requests for additional information. The initial question is transformed into another question, which, if necessary, is transformed further in an analogous way, etc. Consecutive questions clarify the initial problem step by step. The process ends with arriving at a question whose answer is well-known. We coin this type of IQP with the label *ultimate*.

Erotetic calculi enable formal modelling of ultimate IQP. Of course, their area of applicability is restricted to logical problems/questions.

As long as ultimate IQP is concerned, no requests for information are sent, because no additional information is needed. In the case of *distributed* IQP requests for additional information are sent, and questions are transformed into further questions depending on how previous information requests have been fulfilled.

Distributed IQP can be modelled in terms of *erotetic search scenarios*. We address this issue in the consecutive chapters.

---

[8] Cf. Hintikka (1999), Hintikka (2007); see also the next chapter.

Part III

Scenarios

9
E-scenarios

9.1 Erotetic Decomposition Principle

It happens quite often that in order to answer a principal question we have to ask and answer some auxiliary or "operative" questions. Answering them usually amounts to consulting an external source of information (a fellow inquirer, a database, a spouse, etc.). Sometimes new experiments or observations are needed; sometimes the relevant answers belong to our knowledge, but are not taken into account at the moment of asking the principal question. Anyway, auxiliary questions have to be good questions asked at the right time. This idea is intelligible in almost every particular case, but hard to explicate in general. Clearly, answers to auxiliary questions must contribute to the process of finding the right answer to the principal question. Moreover, the order in which the auxiliary questions are asked is important, since the answers received to previously asked questions determine what further questions (if any) are needed. And, last but not least, finding the right answer to an auxiliary question must be less difficult than finding the right answer to the principal question. But these are almost slogans; providing satisfactory explications of the relevant concepts constitutes a serious challenge. What is even more challenging is to provide a formal account of questioning that both makes precise and produces implementations of the following:

(**EDP**) (*Erotetic Decomposition Principle*): *Transform a principal question into auxiliary questions in such a way that: (a) consecutive auxiliary questions are dependent upon previous questions and, possibly, answers to previous auxiliary questions, and (b) once auxiliary questions are resolved, the principal question is resolved as well.*

9.1.1 Interrogative Model of Inquiry

Jaakko Hintikka and his collaborators paid a great deal of attention to the problems mentioned above. In a series of papers published in the eighties and nineties (of the 20th century) the so-called Interrogative Model of Inquiry (IMI) is developed.[1] The concept of *interrogative game* is a central concept of IMI. An interrogative game is played by two parties: an Inquirer and an external source of information, called Nature or Oracle. In the simplest case the aim of a game is to prove a predetermined conclusion, which is an answer to the principal question. In a slightly more complicated case the aim is to prove at least one among previously specified sentences, which are regarded as possible answers to the principal question. Such a game is conceived as consisting of separate games, which are simple games for consecutive answers. Sometimes the aim of an interrogative game is to prove the desideratum of the principal question; a desideratum of a question is, roughly speaking, a proposition which specifies the cognitive state of affairs which the Inquirer wants to be brought about. The situation is more complicated in the context of why-questions, but this does not concern us here. In each case it is assumed that the Inquirer has at his/her disposal some initial premises. The Inquirer can perform moves of the following kinds: (a) *deductive moves*, in which conclusions are drawn from what has already been established; (b) *interrogative moves*, in which auxiliary questions are addressed to a source of information (the answers received are added to the premises and thus can be used in further deductive moves); (c) *"definitory"* moves, in which new concepts are introduced by explicit definitions; (d) *assertoric moves*, in which the conclusion to be proved is strengthened (moves of the third and fourth kind occur only in more sophisticated games). The only restriction imposed on questions which may occur in interrogative moves is that the presuppositions of these questions have to be established, i.e. must be conclusions of some earlier deductive move(s) or belong to the initial premises. In different variants of IMI different restrictions are imposed on the accessibility and reliability of answers to the (potential) operative questions. The Inquirer is free to choose between a deductive move and an interrogative move: he/she can either use the (already obtained) presupposition of a question as a premise in a deductive move or can ask the corresponding question and (possibly) receive new information, which may be used in further derivation(s). The choice between moves as well as the choice between admissible questions is a matter of strategy; interrogative games are called games not in order to use the mathematical results of game theory, but to do justice to the importance of research strategies, modelled in IMI by different questioning strategies.

The following feature of IMI has to be stressed: the deductive moves are the only inferential moves of an interrogative game, and both premises and conclusions of the inferences are declarative sentences/d-wffs. Questions do not perform the roles of premises and conclusions. They are devices by means of which new relevant information comes into play (of course, with the exception of the principal question, which specifies the aim of the game).

---

[1] The papers are collected in Hintikka (1999). For IMI see also, e.g., Hintikka et al. (2002), and Hintikka (2007).

9.2 E-scenarios: intuitions

IEL addresses issues related to the Erotetic Decomposition Principle in many ways. In the case of ultimate internal question processing erotetic rules and Socratic transformations are means of decomposition (see Chapter 8). As for distributed internal question processing, IEL operates with the concept of *erotetic search scenario* (e-scenario for short).

In order to show what e-scenarios are let us start with three stories.

9.2.1 First story

Let us imagine a detective who is trying to keep track of a certain criminal, named Andrew. The detective looks for the answer to the question:

> *What was the destination of Andrew's flight:* (9.1)
> *London, Paris, Moscow, or Rome?*

on the basis of the following pieces of information:

> *The destination was London or Paris if and only if* (9.2)
> *Andrew departed in the morning.*

> *The destination was Rome or Moscow if and only if* (9.3)
> *Andrew departed in the evening.*

> *If Andrew flew by BA, then the destination was* (9.4)
> *neither Paris nor Rome.*

> *If the destination was either London or Moscow,* (0.5)
> *then Andrew flew by BA.*

Clearly, (9.2) – (9.5) are insufficient in order to resolve (9.1). So new information is needed. How can the detective proceed now?

One possibility is to design the following search scenario:

"First, I will ask:

> *When did Andrew depart: in the morning, or in the evening?* (9.6)

If it occurs that Andrew departed in the morning, I will ask:

> *Did Andrew fly by BA in the morning?* (9.7)

If, however, it occurs that Andrew departed in the evening, I will ask:

> *Did Andrew fly by BA in the evening?"* (9.8)

What is the rationale of the scenario? Each direct answer to (9.6), if received, leaves the detective with two "positive" options instead of the initial four. If Andrew departed in the morning, then, by (9.2), the destination was London or Paris; if in the evening, then, by (9.3), either Rome or Moscow was the destination. On the other hand, if the initial options are only London, Paris, Rome, and Moscow, then, by (9.2) and (9.3), Andrew departed either in the morning or in the evening. When the relevant options are reduced to London and Paris, each answer to question (9.6) would give a further reduction. For if Andrew flew by BA in the morning, then, by (9.4), the destination was not Paris and therefore it was London. If Andrew did not fly by BA in the morning, then, by (9.5), London is not the case and hence Paris is. Similarly, when Rome and Moscow are the options to be considered, the affirmative answer to question (9.7), together with premise (9.4), would exclude Moscow and thus give Rome. The negative answer to (9.7), along with premise (9.5), would exclude Rome and hence give Moscow as the outcome.

Figure 9.1 summarizes the case.

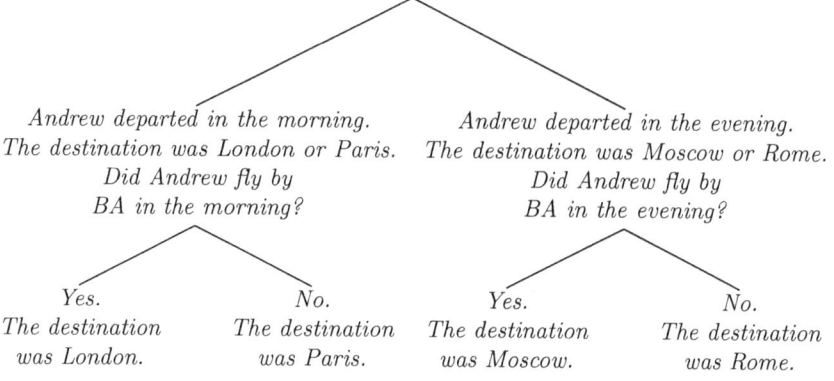

Fig. 9.1. *First story: the search scenario.*

The above scenario consists of four paths or branches. Note that each path satisfies the following conditions:

(1) it begins with the principal question and ends with a direct answer to it;
(2) each declarative sentence involved:[2]
   - is an initial premise, or
   - is a direct answer to an auxiliary question that immediately precedes it on the path, or
   - is entailed by some declarative sentence(s) which occur(s) earlier on the path

---

[2] The short answers "Yes" and "No" are construed as representing the relevant direct answers.

9.2 E-scenarios: intuitions

(3) each auxiliary question involved is implied, in the sense of IEL, by some question and declarative sentence(s) that occur earlier on the path.

As a matter of fact, question (9.6) is implied by question (9.1) on the basis of the (set of) sentences (9.2) and (9.3). Question (9.7) is implied by question (9.1) on the basis of "The destination was either London or Paris", sentence (9.4), and sentence (9.5). Similarly, question (9.8) is implied by question (9.1) on the basis of "The destination was either Rome or Moscow", sentence (9.4), and sentence (9.5).[3]

The scenario as a whole satisfies the following conditions:

(a) no direct answer to the principal question belongs to the set of initial premises;
(b) if an auxiliary question is immediately followed, on a given path, by a direct answer to it, then the scenario contains path(s) on which this question is immediately followed by all the other direct answers to the question; these paths are identical up to the point at which the auxiliary question occurs, but start to differ at the level of answers to the auxiliary question;
(c) only auxiliary questions have more that one immediate successor.

9.2.2 Second story

Let us come back to the detective who was the character in the first story. Now let us imagine that he is looking for the answer to the question:

Where did Andrew leave for: Paris, London or Rome? (9.9)

This time, however, he makes use of the following initial premises:

Andrew left for Paris, London or Rome. (9.10)

If Andrew flew by Air France, then he left for Paris. (9.11)

If Andrew did not fly by Air France, then he did not leave for Rome. (9.12)

Andrew left for London if and only if he flew by BA or Rynair. (9.13)

An option for the detective is to design and then implement a scenario depicted in Figure 9.2.

The scenario fulfils all the conditions (1), (2), (3), (a), (b), and (c) specified in the previous section. In particular, each auxiliary question is an IEL-implied question.[4]

---

[3] Observe that (9.7) is *also* implied by (9.1) on the basis of the set of initial premises supplemented with the answer "Andrew departed in the morning" to question (9.6). Similarly in the case of question (9.8), but with respect to the second answer to question (9.6).
[4] Assuming Classical Logic as the background.

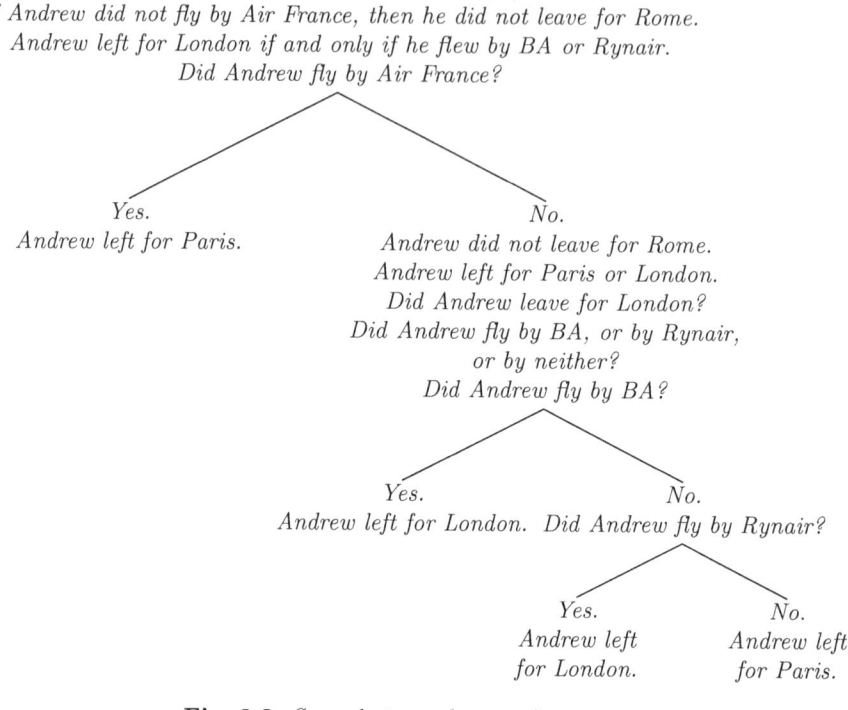

Fig. 9.2. *Second story: the search scenario.*

The question:

$$\text{Did Andrew fly by Air France?} \tag{9.14}$$

is (erotetically) implied by question (9.9) on the basis of the sentences (9.10), (9.11), and (9.12). For if Andrew flew by Air France, then, by (9.11), he left for Paris. If, however, Andrew did not fly by Air France, then, by (9.12), he did not leave for Rome and hence, by (9.10), he left either for Paris or for London.

The question:

$$\text{Did Andrew leave for London?} \tag{9.15}$$

is, obviously, implied by question (9.9) on the basis of "Andrew left for Paris or London."

The consecutive question:

$$\text{Did Andrew fly by BA, or by Rynair, or by neither?} \tag{9.16}$$

is implied by question (9.15) on the basis of (9.13).

Finally, any of the questions:

$$\text{Did Andrew fly by BA?} \tag{9.17}$$

$$\text{Did Andrew fly by Rynair?} \tag{9.18}$$

is implied by question (9.16). To see this it suffices to observe that, taking Classical Logic as the background, we have:

$$\neg A \models \{B, \neg A \wedge \neg B\} \qquad (9.19)$$

$$\neg B \models \{A, \neg A \wedge \neg B\} \qquad (9.20)$$

It is worth emphasizing that, unlike the previous scenario, the scenario depicted in Figure 9.2 has some specific features. First, it is *incomplete* in the sense that there is no path which leads to a certain (direct) answer to the principal question, namely the answer "Andrew left for Rome". Second, the scenario involves auxiliary questions which are not followed by direct answers to them – which do not function as *queries* – but serve as premises for further questions only. In many cases such auxiliary questions are indispensable, since, as we have shown in Chapter 7, sections 7.3.2 and 7.5.2, erotetic implication is not "transitive". As for the scenario considered, in order to arrive at questions (9.17) and (9.18) one needs a transition from question (9.15) to question (9.16) as a necessary step.

9.2.3 Third story

Our third story will be presented in more abstract terms.

Suppose that one is looking for the answer to a question of the form:

Is it the case that p, or is it the case that q, or is it the case that r? (9.21)

Suppose further that it is known that p holds if s holds, and that either q or r holds if $\neg s$ holds. In this situation one arrives at the question:

Is it the case that s? (9.22)

What can happen next? It depends on the epistemic situation. If the request for information will be satisfied by s, the answer p to the initial question is found. If, however, the request will be satisfied by $\neg s$, the initial question transforms into the question:

Is it the case that q, or is it the case that r? (9.23)

Now suppose that it is also known that q holds if, and only if u holds. In this situation one arrives at the question:

Is it the case that u? (9.24)

If this request for information will be satisfied by u, one gets the answer q to the principal question. If the outcome will be $\neg u$, one gets r, since if u does not hold, q does not hold either, and, as $q \vee r$ holds, r must hold.

We have told the story in epistemic terms. Let us now look, however, at the underlying structure displayed in Figure 9.3: [5]

---

[5] We have used language $\mathcal{L}^?_{cpl}$ as the mean of formalization.

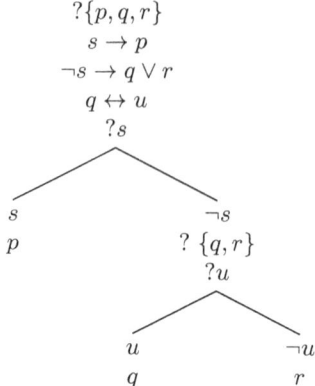

Fig. 9.3. *Third story: the search scenario.*

Figure 9.3 depicts the corresponding e-scenario. It has the form of a tree and is of course completely *domain-unspecific*. Needless to say, the conditions (1) – (3) and (**a**) – (**c**) specified in section 9.2.1 are satisfied.

9.3 E-scenarios: definitions

Erotetic search scenarios (e-scenarios for short) are abstract entities defined in terms of IEL. They can be defined in two equivalent ways: (a) as families of interconnected e-derivations of answers to principal questions (see Wiśniewski (2001), Wiśniewski (2003)), or (b) as labelled trees. In what follows we will present both definitions. We start with the first option, because this step simplifies the metatheoretical considerations.

Our logical basis is just the logical basis of IEL. To be more precise, we assume that we deal with a formal language with questions, $\mathcal{L}$, which satisfies the conditions specified in section 5.4 of Chapter 5. The language is supplemented with a semantics characterized in Chapters 3 and 4, with the concept of admissible partition as the basic one. The conceptual apparatus of IEL introduced in Chapters 6 and 7 is used accordingly. For brevity, we omit the specifications "in $\mathcal{L}$" and "of $\mathcal{L}$".

Terminology and notation. By wffs of $\mathcal{L}$ we mean d-wffs of $\mathcal{L}$ and questions of $\mathcal{L}$. We use $\mathsf{s}_1, \mathsf{s}_2, \ldots, \mathsf{t}_1, \mathsf{t}_2, \ldots$, with or without superscripts, as metalanguage variables for wffs of the language. When we write ds_i, we assume that s_i is a question and we refer to the set of direct answers to s_i.

9.3.1 Erotetic derivations

We need the following auxiliary concept.

Definition 9.1 (*E-derivation*). *A finite sequence* $\mathsf{s} = \mathsf{s}_1, \ldots, \mathsf{s}_n$ *of wffs is an erotetic derivation (e-derivation for short) of a direct answer A to question Q on the basis of a set of d-wffs X iff* $\mathsf{s}_1 = Q$, $\mathsf{s}_n = A$, *and the following conditions hold:*

1. for each question s_k of s such that $k > 1$:
   a. $ds_k \neq dQ$,
   b. s_k is implied by a certain question s_j which precedes s_k in s on the basis of the empty set, or on the basis of a non-empty set of d-wffs such that each element of this set precedes s_k in s, and
   c. s_{k+1} is either a direct answer to s_k or a question;
2. for each d-wff s_i of s:
   a. $s_i \in X$, or
   b. s_i is a direct answer to s_{i-1}, where $s_{i-1} \neq Q$, or
   c. s_i is entailed by a certain non-empty set of d-wffs such that each element of this set precedes s_i in s;

Note that by "precedes" we do not mean "immediately precedes".

If $s = s_1, \ldots, s_n$ is an e-derivation of a direct answer to question Q on the basis of X, each question of s different from Q is called an *auxiliary question* of s, the elements of X are called *initial d-wffs*, and the elements of X which occur in s are called *initial premises* of the e-derivation.

Some comments are in order. An e-derivation is *goal-directed*: it begins with a principal question and ends with a direct answer to the question. The remaining items are either d-wffs or auxiliary questions. Clause (1a) requires that no auxiliary question has the same set of direct answers as the principal question (i.e. question Q). It follows that no auxiliary question is a simple reformulation of the principal one. Clause (1b) amounts to the requirement that each question of an e-derivation different from the principal one must be implied (in the sense of erotetic implication) by some earlier item(s) of the e-derivation. Clause (1c), in turn, requires than an auxiliary question must be immediately succeeded either by a direct answer to it or by a further auxiliary question. According to clause (2), d-wffs may enter e-derivations for three reasons: as initial d-wffs, as direct answers to auxiliary questions (in this case they occur just after the relevant questions), or as consequences of earlier d-wffs. Let us stress that, by the clause (2b), a direct answer to an auxiliary question (but not to the principal question!) may enter an e-derivation even if this answer neither belongs to the set of initial d-wffs X nor is entailed by some earlier item(s) of the derivation.[6]

Let us stress that e-derivations in our sense differ substantially from Hintikka's "interrogative derivations".[7] Harrah sometimes uses the term "e-derivation" (but not "erotetic derivation"), but he understands this term in a way different from ours and uses it for different reasons.[8]

---

[6] But observe that, by clause (2b), a question-answer sequence Q, A is an e-derivation of the answer A if, and only if A is an element of the relevant X, that is, is an initial d-wff. Clause (2b) allows for a "free introduction" of a direct answer only with respect to auxiliary questions.

[7] Cf. the papers included in Hintikka (1999), or the paper Hintikka et al. (2002).

[8] Cf. Harrah (1981), and Harrah (2002). Roughly, e-derivations in Harrah's sense break down compounds (non-atomic wffs made up from e-formulas and possibly d-wffs), which is an important procedure in Harrah's formal theory of message and reply as well as in his General Erotetic Logic.

Queries of e-derivations

There are e-derivations which involve only one question, that is, the principal one. But there are also e-derivations which contain more than one question; we shall call them *interrogatively non-trivial*. Note that interrogatively non-trivial e-derivations must involve *queries*, i.e. auxiliary questions immediately followed by direct answers to them.

Definition 9.2 (*Query of e-derivation*). *A term s_k (where $1 < k < n$) of an e-derivation $s = s_1, \ldots, s_n$ is a query of s if s_k is a question and s_{k+1} is a direct answer to s_k.*

Queries are thus defined syntactically. However, the underlying intuition is: queries of an e-derivation are these auxiliary questions which are "asked and then answered" in it.

We do not require each auxiliary question to be a query. There are e-derivations which involve auxiliary questions that are not queries.

Here are examples of e-derivations:

$$?\{p,q,r\}, s \to p, \neg s \to q \vee r, q \leftrightarrow u, ?s, s, p \qquad (9.25)$$

$$?\{p,q,r\}, s \to p, \neg s \to q \vee r, q \leftrightarrow u, ?s, \neg s, ?\{q,r\}, ?u, u, q \qquad (9.26)$$

$$?\{p,q,r\}, s \to p, \neg s \to q \vee r, q \leftrightarrow u, ?s, \neg s, ?\{q,r\}, ?u, \neg u, r \qquad (9.27)$$

Question $?s$ is the only query of (9.25). Questions $?s$ and $?u$ are queries of (9.26) as well as of (9.27), while question $?\{q,r\}$ is an auxiliary question which is not a query. Clause (1b) of Definition 9.1 is fulfilled because the following hold in $\mathcal{L}_{cpl}^?$:

$$\mathbf{Im}(?\{p,q,r\}, s \to p, \neg s \to q \vee r, ?s) \qquad (9.28)$$

$$\mathbf{Im}(?\{q,r\}, \neg s \to q \vee r, \neg s, q \leftrightarrow u, ?u) \qquad (9.29)$$

Observe that (9.25), (9.26), and (9.27) are sequences of labels of the consecutive branches of the tree displayed in Figure 9.3. This is not by an accident; we will come back to this issue in a moment.

Epistemic justification and minimal error points

Let s be an e-derivation of a direct answer A to question Q on the basis of X. If s involves no query, then each d-wff of s, the answer A included, is true in any admissible partition in which all the d-wffs in X are true. But if s involves a query or queries, the situation may be different, since a d-wff may enter s only because it is a direct answer to a query and there is nothing in Definition 9.1 that prevents us from introducing a false answer. So a mere existence of an e-derivation of a direct answer A to Q on the basis of X is not epistemically relevant; one cannot say that such a derivation is an "interrogative" or "erotetic" *proof* of A from X (even if all the elements of X are properly justified). An e-derivation becomes epistemically relevant when all the declarative premises used in it (answers to queries included!) are properly justified.

Observe that an e-derivation which starts with a sound question, is based on true initial premises, involves at least one query and includes a false d-wff or an unsound question has a *minimal error point* being an index of an answer to a query. One can easily prove the following:

Lemma 9.3. *Let* $s = s_1, \ldots, s_n$ *be an e-derivation of a direct answer* A *to question* Q *on the basis of a set of d-wffs* X. *Assume that* s *involves at least one query. Let* P *be an admissible partition such that all the d-wffs in* X *are true in* P, *and* Q *is sound in* P. *If at least one d-wff of* s *is false in* P, *or at least one question of* s *is unsound in* P, *then there exists an index* i $(1 < i \leq n)$ *such that:*

1. s_i *is false in* P *and* s_i *is a direct answer to* s_{i-1}, *and*
2. *each question which occurs in* s *before* s_i *is sound in* P, *and*
3. *each d-wff which occurs in* s *before* s_i *is true in* P.

Lemma 9.3 will be used in section 9.4 below.

9.3.2 E-scenarios as families of e-derivations

E-scenarios can be defined, first, as families of interconnected e-derivations. Here is the definition.

Definition 9.4 (E-scenario). *A finite family* Σ *of sequences of wffs is an erotetic search scenario (e-scenario for short) for a question* Q *relative to a set of d-wffs* X *iff each element of* Σ *is an e-derivation of a direct answer to* Q *on the basis of* X *and the following conditions hold:*

1. $dQ \cap X = \emptyset$;
2. Σ *contains at least two elements;*
3. *for each element* $s = s_1, \ldots, s_n$ *of* Σ, *for each index* k, *where* $1 \leq k < n$:
   a. *if* s_k *is a question and* s_{k+1} *is a direct answer to* s_k, *then for each direct answer* B *to* s_k: *the family* Σ *contains a certain e-derivation* $s^* = s_1^*, s_2^*, \ldots, s_m^*$ *such that* $s_j = s_j^*$ *for* $j = 1, \ldots, k$, *and* $s_{k+1}^* = B$;
   b. *if* s_k *is a d-wff, or* s_k *is a question and* s_{k+1} *is not a direct answer to* s_k, *then for each e-derivation* $s^* = s_1^*, s_2^*, \ldots, s_m^*$ *in* Σ *such that* $s_j = s_j^*$ *for* $j = 1, \ldots, k$ *we have* $s_{k+1} = s_{k+1}^*$.

The e-derivations which are elements of an erotetic search scenario Σ for Q relative to X will be called *paths* of Σ, the question Q will be called the *principal question* of Σ, and any other question of Σ is called an *auxiliary question* of the e-scenario. The relevant set X will be referred to as the *background*, and the elements of X which occur in Σ will be called *initial premises* of Σ. If a path s of Σ has a direct answer A to Q as its last term, we say that s *leads to* A.

Definition 9.5 (Query of e-scenario). *A query of an e-scenario is a query of a path of the e-scenario.*

A quick look at definitions 9.2 and 9.1 gives the following: a query of an e-scenario is simply the first element of a question-answer pair that occurs on a path of the e-scenario, where the question is an auxiliary one and the answer immediately succeeds the question. Thus each query is a question, but e-scenarios can involve auxiliary questions that are not queries.

Let us now comment on Definition 9.4.

Clause (2) requires an e-scenario to comprise at least two (properly interconnected) e-derivations of direct answer(s) to the principal question. For obvious reasons, it is assumed (and required by clause (1)) that no direct answer to a principal question simply belongs to the background.

Clause (3a) expresses the idea of *fairness with respect to queries*: if A is an answer to a query that immediately succeeds the query on a path $\mathbf{s}$ of an e-scenario Σ, then for *any* direct answer B to the query that is different from A there exists a path $\mathbf{s}^*$ of Σ which is identical with $\mathbf{s}$ to the level of the query, and then has B as the immediate successor of the query. Thus, roughly, for any path and any query on that path there exists a cluster of related paths which share the query and its predecessors, but diverge with respect to the direct answers to the query. Moreover, each direct answer to a query is "used" at some path of the cluster. Or, to put it differently, each direct answer to a query contributes to some path and thus to a derivation of an answer to the principal question: there are no "dead ends".

Clause (3b), in turn, expresses the idea of *regularity*: if $\mathbf{s}_k$ ($k < n$) is a d-wff of a path $\mathbf{s} = \mathbf{s}_1, \ldots, \mathbf{s}_n$, or $\mathbf{s}_k$ is a question of $\mathbf{s}$ that is not a query, then each path which is identical with $\mathbf{s}$ to the level of $\mathbf{s}_k$ has the wff $\mathbf{s}_{k+1}$ as the $k+1$st term. In other words, d-wffs as well as questions that are not queries are "used" within a cluster of related paths in an uniform manner. Hence only queries are "branching points" of e-scenarios. For answers to queries we have:

Corollary 9.6. *Let* $\mathbf{s} = \mathbf{s}_1, \ldots, \mathbf{s}_n$ *be a path of an e-scenario* Σ, *and* $\mathbf{s}_k$ *be a query of* $\mathbf{s}$. *Let* $\mathbf{s}^* = \mathbf{s}_1^*, \ldots, \mathbf{s}_m^*$ *be a path of* Σ *such that* $\mathbf{s}_i^* = \mathbf{s}_i$ *for* $i = 1, \ldots, k+1$. *Then* $\mathbf{s}_{k+2}^* = \mathbf{s}_{k+2}$ *given that* $k+1 < n$. *Moreover, if* $\mathbf{s}_w$ *is the next query of* $\mathbf{s}$, *then* $\mathbf{s}_j^* = \mathbf{s}_j$ *for* $j = k+2, \ldots, w$. *If, however,* $\mathbf{s}_k$ *is the last query of* $\mathbf{s}$, *then* $\mathbf{s}^* = \mathbf{s}$.

Since e-scenarios are supposed to be finite sets, by clause (3a) of Definition 9.4 we get:

Corollary 9.7. *Each query of an e-scenario is a question with a finite number of direct answers.*

Recall that, in view of the general setting adopted, each question has at least two direct answers.

Note that it is permitted that principal questions and auxiliary questions that are not queries have infinitely many direct answers.

We also have:

Corollary 9.8. *Each path of an e-scenario involves at least one query.*

Proof. Assume that there is a path, say, s, of an e-scenario Σ such that s involves no query. As an e-scenario, Σ has at least two elements. Moreover, $dQ \cap X = \emptyset$ and thus the second term of s is not a direct answer to Q. By clause (3b) of Definition 9.4, the second term of any path of Σ is equal to the second term of s. Yet, since s involves no query, then, by clause (3b) again, the consecutive terms of any path of Σ are equal to the corresponding terms of s. Therefore Σ is a singleton set. A contradiction. □

Each e-scenario has the *first query*, shared by all the paths of the scenario. More precisely, the following holds:

Corollary 9.9. *Let Σ be an e-scenario for Q relative to X. There exist an index $k > 1$ and a question Q^* such that:*

1. *Q^* is the k-th term of every path of Σ,*
2. *the k-th term of a path of Σ is a query of the path and hence of Σ,*
3. *k is the least index of a query of Σ.*

Proof. By Corollary 9.8, each path of Σ involves at least one query. Queries are terms of paths and hence any query has an index. For each path there exists the least index of a query of the path. Consider the set $\mathtt{Mqi}_\Sigma$ of the least indices of queries of Σ. Since e-scenarios are finite sets, $\mathtt{Mqi}_\Sigma$ is finite and hence has a minimal element, say, k. In other words, no query of Σ has an index lower than k. Clearly $k > 1$. Let $\mathbf{s} = \mathbf{s}_1, \ldots, \mathbf{s}_n$ be a path of Σ whose least index of a query is k. Assume that $k = 2$. Therefore, by clause (3b) of Definition 9.4, the question $\mathbf{s}_2$ occurs as the second term of any path of Σ, and, by clause (3a) of the definition, is a query of every path of Σ. Now assume that $k > 2$. Suppose that there exists a path, $\mathbf{s}'$, of Σ such that the least index of a query of $\mathbf{s}'$ is k and $\mathbf{s}'_j \neq \mathbf{s}_j$ for some $2 < j \leq k$. For each i, where $1 < i < k$, $\mathbf{s}'_i$ is either a d-wff or an auxiliary question which is not a query. Hence, by clause (3b) of Definition 9.4, $\mathbf{s}'_i = \mathbf{s}_i$ for any $i < k$. It follows, again by the clause (3b), that $\mathbf{s}'_k = \mathbf{s}_k$. A contradiction. Hence the question $\mathbf{s}_k$ is a query of each path of Σ. □

Note that it is not assumed that paths of e-scenarios are sequences without repetitions. Moreover, as we will see, there are cases in which it is intuitive to allow for multiple occurrence(s) of questions and/or d-wffs, although at different paths.

It is obvious that e-scenarios can be displayed in the form of diagrams showing downward trees; the paths of an e-scenario are represented by the branches of a tree (i.e. maximal paths of the tree).

9.3.3 E-scenarios as labelled trees

E-scenarios can also be viewed as *labelled trees*, where the labels are d-wffs and questions. It can be shown that e-scenarios defined as families of e-derivations and as labelled trees stay in a 1-1 correspondence (see Leszczyńska-Jasion (2013)).

Definition 9.10. *A finite labelled tree Φ is an erotetic search scenario for a question Q relative to a set of d-wffs X iff*

1. *the nodes of Φ are labelled by questions and d-wffs; they are called e-nodes and d-nodes, respectively;*
2. *Q labels the root of Φ;*
3. *each leaf of Φ is labelled by a direct answer to Q;*
4. *$\mathsf{d}Q \cap X = \emptyset$;*
5. *for each d-node γ_δ of Φ: if A is the label of γ_δ, then*
   a. *$A \in X$, or*
   b. *$A \in \mathsf{d}Q^*$, where $Q^* \neq Q$ and Q^* labels the immediate predecessor of γ_δ, or*
   c. *$\{B_1, ..., B_n\} \models A$, where B_i $(1 \leq i \leq n)$ labels a d-node of Φ that precedes the d-node γ_δ in Φ;*
6. *each d-node of Φ has at most one immediate successor;*
7. *there exists at least one e-node of Φ which is different from the root;*
8. *for each e-node γ_ε of Φ different from the root: if Q^* is the label of γ_ε, then $\mathsf{d}Q^* \neq \mathsf{d}Q$ and*
   a. *$\mathbf{Im}(Q^{**}, Q^*)$ or $\mathbf{Im}(Q^{**}, B_1, ..., B_n, Q^*)$, where Q^{**} labels an e-node of Φ that precedes γ_ε in Φ and B_i $(1 \leq i \leq n)$ labels a d-node of Φ that precedes γ_ε in Φ, and*
   b. *an immediate successor of γ_ε is either an e-node or is a d-node labelled by a direct answer to the question that labels γ_ε, moreover:*
      - *if an immediate successor of γ_ε is an e-node, it is the only immediate successor of γ_ε,*
      - *if an immediate successor of γ_ε is not an e-node, then for each direct answer to the question that labels γ_ε there exists exactly one immediate successor of γ_ε labelled by the answer.*

A *query* of an e-scenario Φ can be defined as a question that labels an e-node of Φ which is different from the root and whose immediate successor is not an e-node. Paths of e-scenarios can be identified with downward sequences of labels of nodes of branches, that is, sequences having the principal question as the first term and direct answers to the question as last terms.

In what follows we will, primarily, construe e-scenarios as families of e-derivations, that is, according to Definition 9.4.

9.4 The Golden Path Theorem

The following theorem characterizes an interesting property of e-scenarios:

Theorem 9.11 (*Golden Path Theorem*). *Let Σ be an e-scenario for a question Q relative to a set of d-wffs X. Assume that Q is sound in an admissible partition P, and all the d-wffs in X are true in P. The e-scenario Σ contains at least one path $\mathbf{s}$ such that:*

1. *each d-wff of $\mathbf{s}$ is true in P,*
2. *each question of $\mathbf{s}$ is sound in P, and*
3. *$\mathbf{s}$ leads to a direct answer to Q which is true in P.*

Proof. Let P be an arbitrary but fixed admissible partition such that Q is sound in P and all the d-wffs in X are true in P. In what follows by "sound" we mean "sound in P", by "true" we mean "true in P", and similarly for falsity.

By Corollary 9.8 each path of Σ involves at least one query. Suppose that each path of Σ is fallacious, i.e. at least one d-wff on it is false or at least one question on it is not sound. Thus by Lemma 9.3 each path of Σ has a minimal error point, that is, for each path **s** of Σ there exists an index i such that the i-th term of **s** is a false direct answer to a question which is the $(i-1)$-st term of **s**, and each d-wff/question which occurs in **s** before its i-th term is true/sound. Let us consider the set Mep_Σ of minimal error points of all the paths of Σ (Mep_Φ is of course the set of indices whose elements fulfil the conditions specified above). Since Σ is a finite family of e-derivations, the set Mep_Σ is finite and thus has a maximal element. So there exist an index k (a maximal element of Mep_Σ) and an e-derivation $\mathbf{s} = \mathbf{s}_1, \ldots, \mathbf{s}_n$ in Σ such that for each $j < k$, $\mathbf{s}_j$ is true/sound, $\mathbf{s}_k$ is a false direct answer to $\mathbf{s}_{k-1}$, and Σ contains no e-derivation whose k-th term is true/sound and all the previous terms are true/sound. On the other hand, since $\mathbf{s}_{k-1}$ is sound, then at least one direct answer to it is true. Thus, by clause (3a) of Definition 9.4, Σ contains an e-derivation $\mathbf{s}^* = \mathbf{s}_1^*, \ldots, \mathbf{s}_m^*$ such that $\mathbf{s}_j^* = \mathbf{s}_j$ for $j = 1, \ldots, k-1$ and $\mathbf{s}_k^*$ is a true direct answer to $\mathbf{s}_{k-1}^*$ (i.e. to $\mathbf{s}_{k-1}$). Since, by assumption, each path of Σ is fallacious, it follows that the minimal error point of $\mathbf{s}^*$ is greater than the maximal element of the set Mep_Σ of minimal error points of all the paths of Σ. We arrive at a contradiction. Therefore at least one path of Σ is not fallacious, that is, each question which occurs on it is sound and each d-wff of it is true. On the other hand, the last d-wff of a path of Σ is a direct answer to Q. □

Theorem 9.11 says that an e-scenario contains at least one *golden path* if the principal question is sound and all the d-wffs in the background are true. Of course, by a golden path of an e-scenario we mean a path which involves only sound questions and only true d-wffs, and which leads to a true direct answer to the principal question of the scenario.

Remarks. Auxiliary questions of e-scenarios are required to be IEL-implied. However, the only property of erotetic implication that is important for the proof of the Golden Path Theorem is the "transmission of soundness/truth into soundness" feature. The second basic feature of erotetic implication, the "open-minded cognitive usefulness" property, safeguards local relevance of consecutive auxiliary questions, while their global relevance is due to the fact that each path of an e-scenario ends with a direct answer to the principal question.

9.5 A pragmatic account of e-scenarios

E-scenarios are abstract entities defined in terms of IEL. But, from a pragmatic point of view, an e-scenario provides us with conditional instructions which tell what auxiliary questions should be asked and when they should be asked. Queries of e-scenarios can be viewed as requests for information. An e-scenario shows what is the next advisable query if the information request of the previous query has been satisfied in such–and–such way. What is important, an e-scenario does this with regard to any possible way of satisfying the request, where the ways are determined by direct answers to the question which functions as a query. Moreover, an e-scenario behaves in this manner in the case of every query of the e-scenario. Thus the e-scenarios approach transcends the common schema of "production of a sequence of questions and affirmations", and the fact that information requests can be satisfied in one way or another is treated seriously.

Let us illustrate this by examples. Take the e-scenario displayed in Figure 9.4:[9]

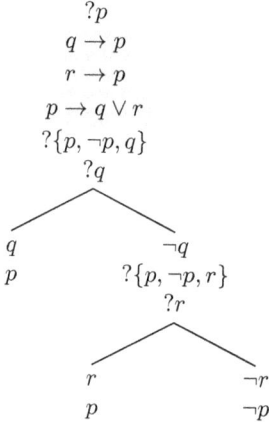

Fig. 9.4. *An example of e-scenario for ?p.*

In order to improve readability initial premises are shadowed, here and below.

Let us comment on the e-scenario depicted in Figure 9.4. Either of q and r constitutes a sufficient condition for p, and their disjunction constitutes a necessary condition for p. So it is advisable to ask first if one of them holds. If it does, there is no need for a further question, and the initial issue is resolved affirmatively. If not, it is advisable to ask whether the other holds. If it does, the initial issue is resolved affirmatively again. If not, the issue is, finally, resolved negatively.[10]

Now take the e-scenario displayed in Figure 9.5:

---

[9] Again, we use the language $\mathcal{L}^?_{cpl}$ (see Chapter 2, section 2.4.1) as the mean of formalization.

[10] Note that neither $?\{p, \neg p, q\}$ nor $?\{p, \neg p, r\}$ is a query. However, they are necessary in the IEL-grounded transitions which lead to queries (see section 7.5.2 of Chapter 7). The e-scenarios depicted in Figures 9.4 and 9.5 are based on the examples (7.30), (7.16), (7.31), (7.17) of erotetic implication given in sections 7.5.1 and 7.5.2 of Chapter 7.

9.5 A pragmatic account of e-scenarios   119

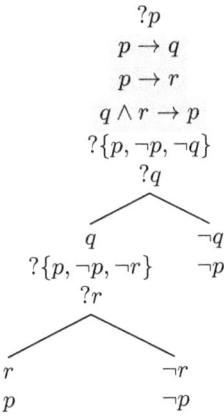

Fig. 9.5. *Another example of e-scenario for ?p.*

This time q and r are necessary conditions for p, and their conjunction constitutes a sufficient condition for p. It is, first, advisable to ask whether q holds. If q holds, it is advisable to ask whether r holds as well. If it does, the problem is resolved affirmatively; otherwise it is resolved negatively. But if the answer to the first query, that is, to $?q$, is negative, the initial problem is already resolved negatively and there is no reason for asking the remaining query, i.e. $?r$. Our next example, presented in Figure 9.6, is, as a matter of fact, a formalization of the content of Figure 9.2, which summarized our "second story" (see section 9.2.2).

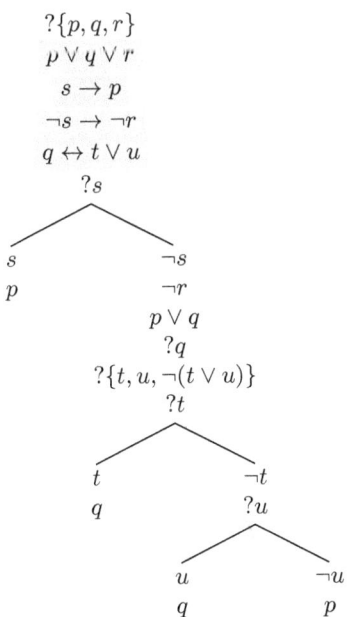

Fig. 9.6. *An example of e-scenario for $?\{p,q,r\}$.*

The following comments on Figure 9.6 are in order. The principal question is ?$\{p, q, r\}$, so the task is accomplished when any of p, q, r is established. Given the initial premises, it is advisable to ask first whether s holds. If it holds, p is the case and there is no need for any further queries. If, however, $\neg s$ is the case, it is advisable to ask whether t holds. If t holds, q is the case and no further query is needed. If, however, $\neg t$ holds, it is advisable to ask whether u holds. If u holds, q is the case; otherwise p is the case.

Let us now analyse some e-scenarios formulated in the language $\mathcal{L}^?_{fom}$ (cf. section 2.4.3 of Chapter 2). P, R, T are supposed to be (object-level!) predicates of $\mathcal{L}^?_{fom}$, a, b, c are individual constants of the language, and x is an object-level individual variable.

In constructing the e-scenarios depicted in Figures 9.7 and 9.8 we made use of examples (7.50) and (7.35) of erotetic implication presented in section 7.5.2 of Chapter 7.

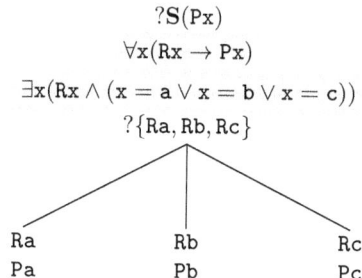

Fig. 9.7. *An example of e-scenario for* ?S(Px).

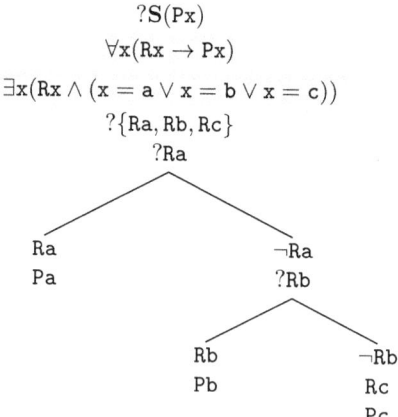

Fig. 9.8. *Another example of an e-scenario for* ?S(Px) .

According to the e-scenario displayed in Figure 9.7, if one is looking for an object having a property P, it is known that objects having property R have the property P, and it is known that at least one of the objects (designated by) a, b, c has the property R, it is advisable to ask which of a, b, c has the property R. The e-scenario displayed in Figure 9.8 pertains to the same principal question and relies upon identical initial premises, but is more sophisticated. It shows that it is advisable to ask, first, whether object a has the property R. If so, it occurs that object a has the property P and no further query is needed (since the principal question is an *existential* which-question). If the answer to the first query is negative, it is advisable to ask whether Rb holds. If the answer is affirmative, we have Pb. If, however, the answer is negative, we get Pc. There is no need for asking, in addition, whether Rc holds.

The e-scenario depicted in Figure 9.7 involves exactly one query which is also the only auxiliary question of the e-scenario. The query is regularly implied by the principal question on the basis of the initial premises. Note that when we know that a question Q_1 with a finite number of direct answers is regularly implied by a question Q on the basis of a finite set of d-wffs X, and the set does not include a direct answer to Q, we can always build an e-scenario whose principal question is Q and the only query is Q_1. Figure 9.9 presents another example of an e-scenario of this kind.[11] The pragmatic information provided

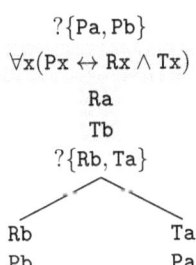

Fig. 9.9. *An example of e-scenario for* ?{Pa, Pb}.

is: given the premises, it is advisable to ask which of the following, Rb or Ta, holds.

Finally, let us consider the e-scenario displayed in Figure 9.10. It is is very similar to, but not identical with that depicted in Figure 9.6. The difference lies in the position of the initial premise $q \leftrightarrow t \vee u$: now it occurs after the first query and is not a term of the leftmost branch.[12] The new pragmatic information provided is the following: if you received the negative answer to the first query and, *in addition*, it is known that q holds if, and only if either t or u holds, it is advisable to ask whether t is the case. The first query does not

---

[11] We make use of (7.34) from section 7.5.2 of Chapter 7.
[12] However, erotetic implication is still retained, since we have $\mathbf{Im}(?q, q \leftrightarrow t \vee u, ?\{t, u, \neg(t \vee u)\})$.

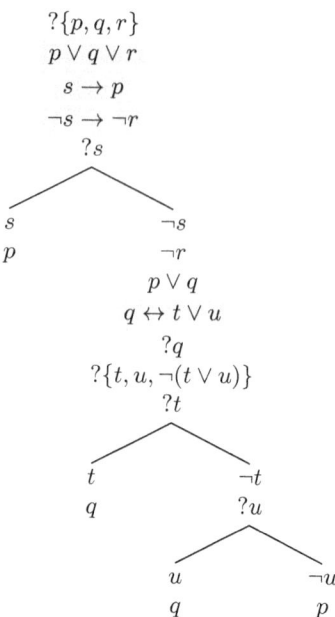

Fig. 9.10. *Another example of e-scenario for* $?\{p,q,r\}$.

rely on this additional information, but the consecutive queries are dependent upon it.

A comment. Neither Definition 9.4 nor Definition 9.10 of e-scenarios require all the initial premises to occur before the first query. E-scenarios which have this property are said to be in the *canonical form*. As we have seen, e-scenarios which are not in the canonical form can carry more pragmatic information than these in the canonical form.

9.5.1 Compression and conciseness

Paths of e-scenarios are e-derivations and thus sequences of questions and d-wffs. Recall that, according to Definition 9.1, a d-wff can enter a path for three possible reasons: (a) as an initial premise taken from the background, (b) as a (direct) answer to a query, or (c) as a consequence of a d-wff or d-wffs introduced in the first and/or second manner. Note that d-wffs fulfilling the condition (c) are, with the exception of last terms of paths, redundant. For, if the last term of a path – a direct answer to the principal question – is entailed by d-wffs introduced due to (a), (b), and (c), the answer is also entailed by the respective d-wffs introduced according to (a) and (b), that is, by initial premises and answers to queries. Moreover, erotetic implications would be retained even if all the d-wffs introduced according to (c) were deleted. This is due to:

Corollary 9.12. *If* $\mathbf{Im}(Q, X \cup \{C\}, Q_1)$ *and* $X \models C$, *then* $\mathbf{Im}(Q, X, Q_1)$.

Proof. It suffices to observe that when $X \models C$, then for each admissible partition $\mathsf{P} = \langle \mathsf{T_P}, \mathsf{U_P} \rangle$ we have $X \subset \mathsf{T_P}$ iff $X \cup \{C\} \subset \mathsf{T_P}$. □

9.5 A pragmatic account of e-scenarios   123

Hence if we *compress* an e-scenario, that is, we delete from it all the d-wffs which are neither initial premises, nor answers to queries, nor last terms of paths, we will receive an e-scenario.

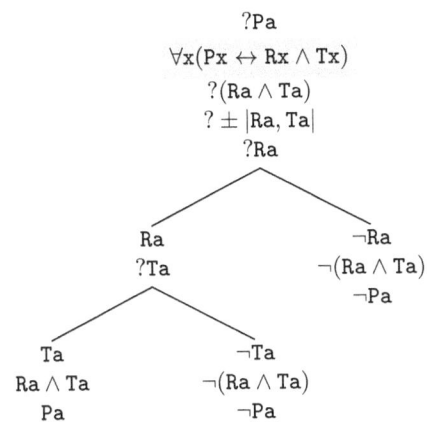

Fig. 9.11. *An example of e-scenario for ?Pa.*

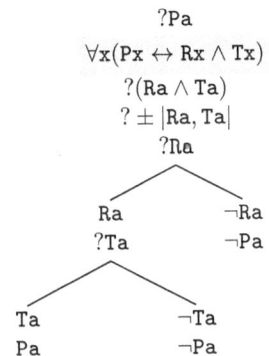

Fig. 9.12. *The compressed counterpart of the e-scenario presented in Figure 9.11.*

Let us illustrate this by an example. Figure 9.11 presents an e-scenario that involves "intermediate" d-wffs.[13] Figure 9.12 shows its compressed counterpart.

Note that compressed counterparts of e-scenarios carry analogous information on preconditions of queries as the original e-scenarios.

Let us introduce the following concept:

Definition 9.13 (*Concise e-scenario*). *An e-scenario Σ for Q relative to X is concise iff each d-wff that is a term of a path of Σ is: (a) an element of X,*

---

[13] Namely, Ra ∧ Ta and ¬(Ra ∧ Ta). Erotetic implication is retained due to (7.32), (7.13) and (7.8) (see section 7.5.2 of Chapter 7).

or (b) a direct answer to a query of the path that occurs immediately after the query on the path, or (c) a direct answer to Q which is the last term of the path.

Compressed counterparts of e-scenarios are concise. There exist e-scenarios which are already concise and thus are not subjected to a compression. The e-scenarios displayed in Figures 9.3, 9.4, 9.5, 9.7, 9.9 are of this kind.

9.5.2 Imperative counterparts of e-scenarios

From a pragmatic point of view, auxiliary questions that are not queries are irrelevant. If we delete them from an e-scenario, we receive the *imperative counterpart* of the initial e-scenario. Figure 9.13 shows the imperative counterpart of the e-scenario displayed in Figure 9.11. Figure 9.14, in turn, depicts the imperative counterpart of e-scenario displayed in figure 9.5.

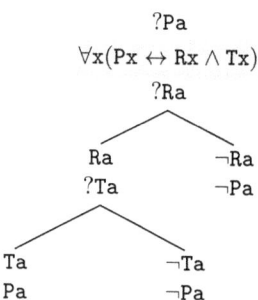

Fig. 9.13. *The imperative counterpart of the e-scenario presented in Figure 9.12.*

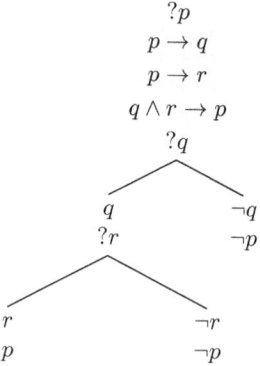

Fig. 9.14. *The imperative counterpart of the e-scenario displayed in Figure 9.5.*

If each auxiliary question of an e-scenario is a query, its imperative counterpart can be identified with the e-scenario itself. Thus Figures 9.7 and 9.9

can be also regarded as displaying the imperative counterparts of the relevant e-scenarios.

Since erotetic implication is not "transitive", the imperative counterpart of an e-scenario need not be an e-scenario itself. But imperative counterparts are carriers of the same pragmatic information as the initial e-scenarios: they characterize preconditions of consecutive queries, in particular preconditions being answers to previous queries.

Although imperative counterparts of e-scenarios need not be e-scenarios, they retain the golden path property. One can easily prove that the analogue of the Golden Path Theorem holds for imperative counterparts.

10

Some Special Kinds of E-scenarios

In this chapter we distinguish some categories of e-scenarios which seem most important from the point of view of possible applications.

10.1 Complete and incomplete e-scenarios

The paths of an e-scenario end with direct answers to the principal question. However, Definition 9.4 does not require each direct answer to the principal question to be the endpoint of a path. Thus we can distinguish between complete and incomplete e-scenarios.

Definition 10.1 (*Complete e-scenario*). *An e-scenario Σ for Q relative to X is complete if each direct answer to Q is the last term of a path of Σ; otherwise Σ is incomplete.*

Speaking in terms of (labelled) trees: if each direct answer to Q is (a label of) a certain leaf, the e-scenario is complete; if the leaves are (labelled by) only some, but not all direct answers to Q, the e-scenario is incomplete.

The e-scenarios depicted in Figures 9.10, 9.8, 9.7, 9.6, 9.2 are incomplete, whereas the remaining e-scenarios presented above are complete.

Since e-scenarios are finite sets of e-derivations/finite labelled trees, we have:

Corollary 10.2. *The principal question of a complete e-scenario has a finite number of direct answers.*

Note that even a complete e-scenario can have different paths which lead to the same direct answer to the principal question. The e-scenarios presented in Figures 9.12, 9.11, 9.5, 9.4 are cases in point here.

10.2 Pure e-scenarios and standard e-scenarios

Definition 9.4 of e-scenarios (as well as Definition 9.10) allows both the background X and sets of initial premises to be empty. In order to distinguish e-scenarios which do not rely upon any initial premise(s) we introduce:

Definition 10.3 (*Pure e-scenario*). *A pure e-scenario is an e-scenario which does not involve any initial premise.*

Figure 10.1 presents an example of a pure e-scenario.

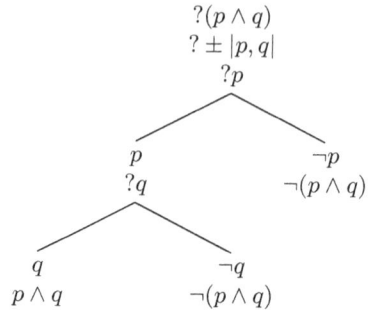

Fig. 10.1. *An example of a pure e-scenario.*

10.2.1 Standard e-scenarios for logical constants. The case of Classical Logic

Let us now be more specific. Assume that questions and d-wffs involved belong, syntactically and semantically, either to the language $\mathcal{L}^?_{cpl}$ or to the language $\mathcal{L}^?_{fom}$.

A *simple yes-no* question has the form:

$$?\{A, \neg A\}$$

and is abbreviated as:

$$?A$$

We say that question of the form $?\{A, \neg A\}$ *is based on* the d-wff A.

Recall the following facts about erotetic implication (in $\mathcal{L}^?_{cpl}$ as well as in $\mathcal{L}^?_{fom}$):

$$\mathbf{Im}(?\neg A, ?A)$$

$$\mathbf{Im}(?(A \otimes B), ?\pm|A, B|)$$

where $\otimes$ is any of the connectives: $\wedge, \vee, \rightarrow, \leftrightarrow$.

$$\mathbf{Im}(?\pm|A, B|, ?A)$$

$$\mathbf{Im}(?\pm|A, B|, ?B)$$

Figures 10.2, 10.3, 10.4, 10.5, 10.6 present schemas of standard e-scenarios for simple yes-no questions whose affirmative answers are compound d-wffs. They are pure e-scenarios. Note that the queries involved are also simple yes-no questions and that their affirmative answers are proper subformulas of affirmative answers to principal questions. In other words, the queries are based on proper subformulas of the d-wff on which the principal question is based.

10.2 Pure e-scenarios and standard e-scenarios   129

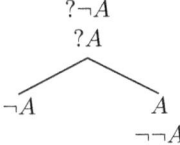

Fig. 10.2. *A schema of the standard e-scenario for negation.*

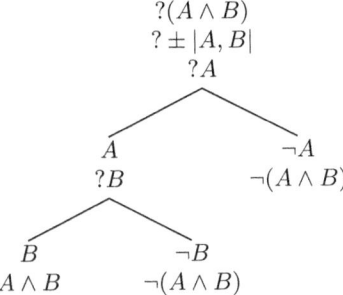

Fig. 10.3. *A schema of the standard e-scenario for conjunction.*

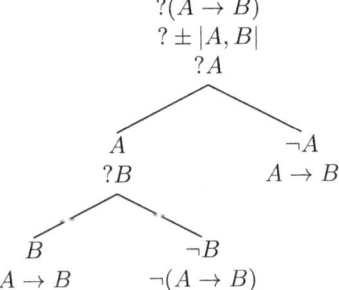

Fig. 10.4. *A schema of the standard e-scenario for implication.*

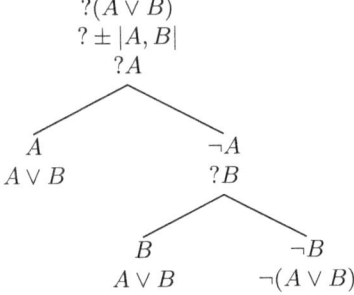

Fig. 10.5. *A schema of the standard e-scenario for disjunction.*

130   10 Some Special Kinds of E-scenarios

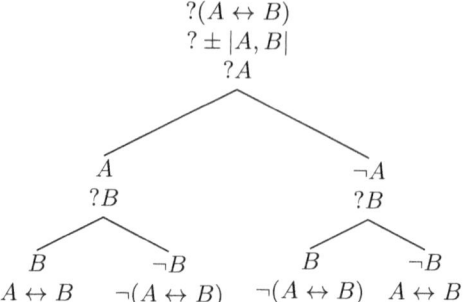

Fig. 10.6. *A schema of the standard e-scenario for equivalence.*

What about quantified d-wffs? Figures 10.7 and 10.8 depict schemas of the appropriate standard e-scenarios which, again, are pure e-scenarios.

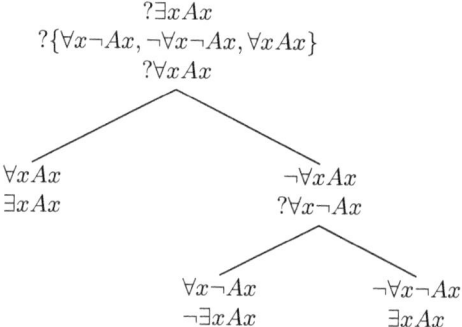

Fig. 10.7. *A schema of the standard e-scenario for an existential quantifier.*

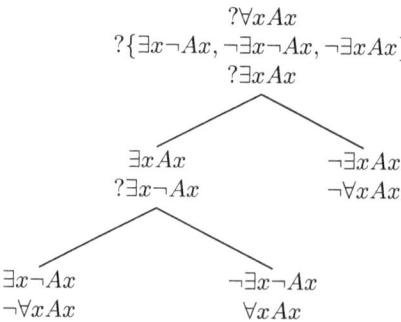

Fig. 10.8. *A schema of the standard e-scenario for a universal quantifier.*

10.2 Pure e-scenarios and standard e-scenarios

Note that e-scenarios falling under the schemas presented by Figures 10.2 – 10.8 are complete. In constructing the e-scenarios (10.7) and (10.8) we made use of the following facts about erotetic implication in $\mathcal{L}^?_{fom}$:

$$\mathbf{Im}(?\exists x Ax, ?\{\forall x \neg Ax, \neg \forall x \neg Ax, \forall x Ax\})$$

$$\mathbf{Im}(?\{\forall x \neg Ax, \neg \forall x \neg Ax, \forall x Ax\}, ?\forall x Ax)$$

$$\mathbf{Im}(?\forall x Ax, ?\{\exists x \neg Ax, \neg \exists x \neg Ax, \neg \exists x Ax\})$$

$$\mathbf{Im}(?\{\exists x \neg Ax, \neg \exists x \neg Ax, \neg \exists x Ax\}, ?\exists x Ax)$$

10.2.2 Standard decomposition e-scenarios

Let us now consider a question (of $\mathcal{L}^?_{cpl}$ or of $\mathcal{L}^?_{fom}$) having the form:

$$?\{A_1, \ldots, A_n\} \tag{10.1}$$

Assume that (10.1) is a safe question.[1] Thus we have:

$$\emptyset \models \{A_1, \ldots, A_n\} \tag{10.2}$$

$$\{\neg A_1, \ldots, \neg A_{n-1}\} \models A_n \tag{10.3}$$

$$\neg A_i \models \{A_1, \ldots, A_n\} \setminus \{A_i\} \tag{10.4}$$

$$\mathbf{Im}(?\{A_1, \ldots, A_n\}, ?A_i) \tag{10.5}$$

for $i = 1, \ldots, n$. Now assume that (10.1) is not a simple yes-no question. Figure 10.9 presents a schema of a pure e-scenario which is applicable to the case. Note that the queries are simple yes-no questions based on direct answers to the principal question.[2]

The situation gets slightly more complicated when (10.1) is not a safe question. In such a case one cannot rely on (10.2) – (10.5). However, the following hold:

$$\{A_1 \vee \ldots \vee A_n, \neg A_1, \ldots, \neg A_{n-1}\} \models A_n \tag{10.6}$$

$$\mathbf{Im}(?\{A_1, \ldots, A_n\}, A_1 \vee \ldots \vee A_n, ?A_i) \tag{10.7}$$

for $i = 1, \ldots, n$.

Figure 10.10 presents a schema of *a standard decomposition e-scenario* for a risky question of the form (10.1) based on (the set made of) $A_1 \vee \ldots \vee A_n$. The e-scenario is not pure. However, the queries are still yes-no questions based on direct answers to the principal question.

The disjunction of all the direct answers to (10.1), $A_1 \vee \ldots \vee A_n$, is a prospective presupposition of (10.1). As a matter of fact, the disjunction occurring in Figure 10.10 can be replaced by any prospective presupposition of the question, in particular by a disjunction of all the direct answers in which the disjuncts are ordered differently.

---

[1] See Definition 4.2 in Chapter 4.
[2] But not on all of them; A_n is not questioned. By the way, this warrants that when (10.1) is of the form $?\{\neg A, A\}$, Figure 10.9 still depicts an e-scenario; otherwise clause (1a) of Definition 9.1 would be violated.

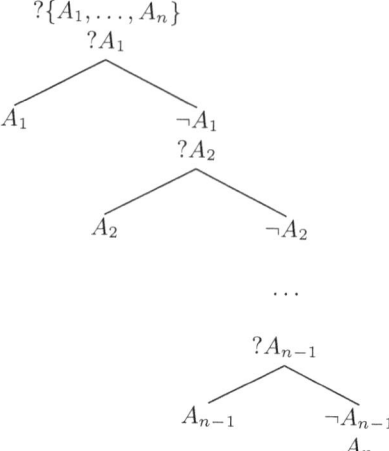

Fig. 10.9. *A schema of the standard decomposition e-scenario for a safe whether-question that is not a simple yes-no question.*

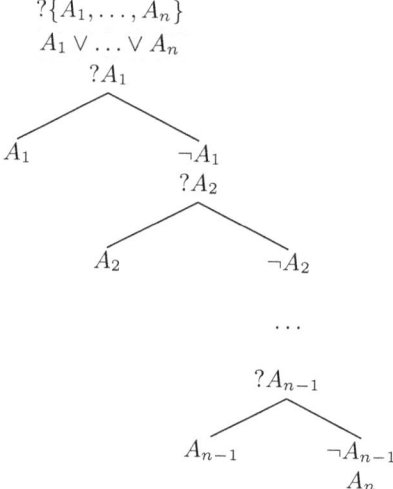

Fig. 10.10. *A schema of the standard decomposition e-scenario for a risky whether-question.*

What about questions which have an infinite number of direct answers? According to Corollary 10.2, there are no complete e-scenarios for them. But still, incomplete e-scenarios for such questions are possible. As for decomposition, we need as an initial premise a d-wff which warrants that a true direct answer to the principal question belongs to a given finite subset of the set of direct answers. To be more precise, let Q be a question with an infinite number of direct answers, and $B_Q^{A_1,\ldots,A_k}$ be a d-wff such that:

$$B_Q^{A_1,\ldots,A_k} \mathrel{\Vert\!=} \{A_1,\ldots,A_k\} \tag{10.8}$$

where $\{A_1,\ldots,A_k\} \subset dQ$, $k > 1$ and $B_Q^{A_1,\ldots,A_k} \notin dQ$. Thus the following hold:

$$\{B_Q^{A_1,\ldots,A_k}, \neg A_1, \ldots, \neg A_{k-1}\} \models A_k \tag{10.9}$$

$$\{B_Q^{A_1,\ldots,A_k}, \neg A_i\} \,\|\!\!\models dQ \setminus \{A_i\} \tag{10.10}$$

for $i = 1,\ldots,k$, and therefore:

$$\mathbf{Im}(Q, B_Q^{A_1,\ldots,A_k}, ?A_i) \tag{10.11}$$

where $1 \leq i \leq k$. Given the above assumptions, Figure 10.11 presents a schema of a decomposition e-scenario for Q with the d-wff $B_Q^{A_1,\ldots,A_k}$ as the only initial premise.

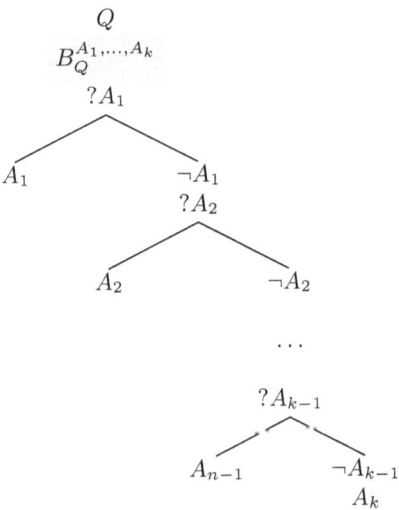

Fig. 10.11. *A schema of a decomposition e-scenario for a question with an infinite number of direct answers.*

Remarks. What is crucial for the results sketched above is not the form of questions, but the fact that logical constants are understood as in Classical Logic. We use the languages $\mathcal{L}_{cpl}^?$ and $\mathcal{L}_{fom}^?$ for illustrative purposes only. The picture can change when connectives and/or quantifiers are construed non-classically.

We are not claiming that the decomposition e-scenarios described in this section are always best from a pragmatic point of view.

10.3 Information-picking e-scenarios

The next category to be distinguished are *information-picking* e-scenarios.

134   10 Some Special Kinds of E-scenarios

We need some auxiliary notions.

We say that a question Q is *informative relative to* a set of d-wffs Z iff no direct answer to Q is entailed by Z.

Let $\mathbf{s} = \mathbf{s}_1, \ldots, \mathbf{s}_n$ be a path of an e-scenario Σ. We define:

$$dec_{\mathbf{s}}^{<}(\mathbf{s}_k) = \{\mathbf{s}_j : j < k \text{ and } \mathbf{s}_j \text{ is a d-wff}\}$$

Thus $dec_{\mathbf{s}}^{<}(\mathbf{s}_k)$ is the set of d-wffs made up of terms of $\mathbf{s}$ that occur in $\mathbf{s}$ before $\mathbf{s}_k$. Note that $dec_{\mathbf{s}}^{<}(\mathbf{s}_k)$ can be empty.

Definition 10.4 (*Information-picking e-scenario*). *An e-scenario Σ for Q relative to X is information-picking iff:*

1. *Q is informative relative to the set of initial premises of Σ, and*
2. *for each path $\mathbf{s} = \mathbf{s}_1, \ldots, \mathbf{s}_n$ of Σ:*
    a. *if $\mathbf{s}_k$ is a query of $\mathbf{s}$, then $\mathbf{s}_k$ is informative relative to $dec_{\mathbf{s}}^{<}(\mathbf{s}_k)$, and $\mathbf{s}_{k+1} \notin \mathrm{d}Q$,*
    b. *$dec_{\mathbf{s}}^{<}(\mathbf{s}_n)$ entails $\mathbf{s}_n$.*

The underlying intuitions are the following. First, due to the clause (1), the principal question cannot be resolved by means of the initial premises only and thus new information is needed. Second, new information is collected gradually: by (2a), no query can be resolved by means of information received so far together with the initial premises used and hence a direct answer to a query brings in new information. Moreover, no direct answer to a query is itself a direct answer to the principal question. Third, clause (2b) amounts to the following: the last term of a path – a direct answer to the principal question – is to be entailed by the initial premises of Σ that occur on the path supplemented with the answers to queries that occur on the path.

All the exemplary e-scenarios presented in Chapter 9 are information-picking.

Let us stress that clause (1) of Definition 10.4 pertains to the set of initial premises of an e-scenario and thus not necessarily to the background of the scenario. When we have an e-scenario for Q relative to X, there is nothing in Definition 9.4 (and similarly in Definition 9.10) that forces all the elements of X to function as initial premises (i.e. to occur at a path) of the e-scenario. In other words, the background and the set of initial premises of an e-scenario need not be equal: the former can be a superset of the latter.[3] So if Σ is an e-scenario for Q relative to X and Σ is information-picking, question Q is informative relative to the set of initial premises of Σ, but not necessarily relative to the whole X or a superset of X. Hence the following cognitive situation can be modelled in terms of information-picking e-scenarios. Let Σ be an information-picking e-scenario for Q relative to X such that the set of initial premises of Σ, say, X^*, is a proper subset of X. Suppose that Q is not informative relative to X,

---

[3] Looking from the purely formal point of view, an e-scenario for Q relative to X is an e-scenario for Q relative to any superset of X that is disjoint with $\mathrm{d}Q$. On the other hand, an e-scenario Σ for Q relative to X is also an e-scenario for Q relative to the set of initial premises of Σ.

that is, to speak generally, Q can be resolved by means of X. What Σ does, pragmatically, is to show what information one should attempt to retrieve from X in order to solve the problem expressed by Q.

Information-picking e-scenarios seem to constitute an interesting class from the point of view of possible applications. This is not to say, however, that only such e-scenarios are relevant in this respect. For instance, let us consider the e-scenario presented by Figure 10.12:

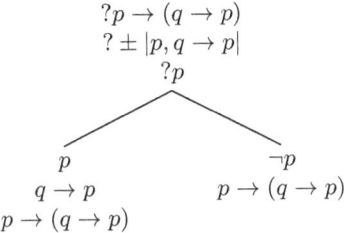

Fig. 10.12. *A pure e-scenario from which a synthetic tableau can be extracted.*

Figure 10.12 depicts a pure e-scenario. It fulfils the clauses (2a) and (2b) of Definition 10.4, but is not information-picking since the affirmative answer to the principal question is a CPL-valid formula and thus clause (1) of the definition does not hold. However, the e-scenario seems to show the following: whatever the logical values of p and q are, the affirmative answer to the principal question is the case. Figure 10.13 presents a more sophisticated e-scenario with analogous properties.

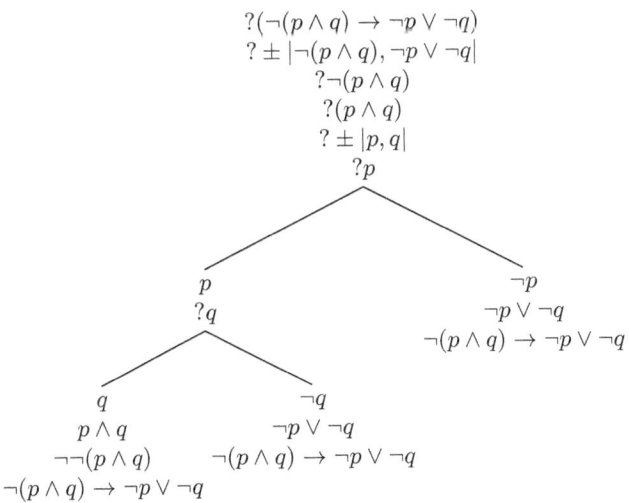

Fig. 10.13. *A pure e-scenario from which a synthetic tableau can be extracted.*

When we delete questions from paths of the e-scenario displayed in Figure 10.13, we will get the sequences (10.12), (10.13), (10.14) of CPL-formulas:

$$p, q, p \wedge q, \neg\neg(p \wedge q), \neg(p \wedge q) \to \neg p \vee \neg q \qquad (10.12)$$

$$p, \neg q, \neg p \vee \neg q, \neg(p \wedge q) \to \neg p \vee \neg q \qquad (10.13)$$

$$\neg p, \neg p \vee \neg q, \neg(p \wedge q) \to \neg p \vee \neg q \qquad (10.14)$$

Each of the sequences (10.12), (10.13), (10.14) "synthesizes" the affirmative answer to the principal question from its subformulas and/or negations of its subformulas. The sequences are interconnected, generally speaking, in the way characterized by Figure 10.13. They, jointly, constitute a *synthetic tableau* for the affirmative answer, which, let us recall, is a CPL-valid formula. The concept of a synthetic tableau, as well as methods and rules of building such tableaux, can be made precise. As an outcome we get a proof method, the Synthetic Tableaux Method, proposed and developed by Mariusz Urbański.[4]

---

[4] Cf. Urbański (2001a), Urbański (2001b), Urbański (2002a), Urbański (2002b).

11

Operations on E-scenarios

Where do e-scenarios come from? The simplest answer is: they simply exist, just as other logical objects do. If one is not pleased with this Platonic answer, the next plausible answer is: they are constructed, and in order to construct them one needs a logic of questions which determines erotetic implication, and a logic of declaratives that determines entailment.

An interesting feature of e-scenarios is that some of them can be obtained from already existing/constructed e-scenarios. In this chapter we will describe two operations which lead, under some conditions, from e-scenarios to e-scenarios.

11.1 Embedding

The first operation can be called *embedding*. Let us start with examples.

Example 11.1. Consider the e-scenarios displayed in Figures 11.1 and 11.2.

Question ?(Ra ∧ Ta) is a query of the e-scenario depicted in Figure 11.1 and is the principal question of the (complete!) e-scenario displayed in Figure 11.2. We embed the latter e-scenario into the former. The result is the e-scenario displayed in Figure 11.3 (already presented in section 9.5.1 of Chapter 9).

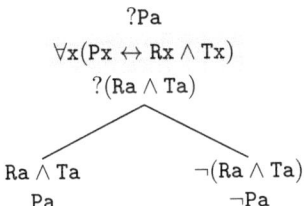

Fig. 11.1. *The e-scenario into which an e-scenario will be embedded.*

Some comments are in order. Question ?(Ra ∧ Ta) is a query of the e-scenario displayed in Figure 11.1. The e-scenario "shows" that once Ra ∧ Ta is the case,

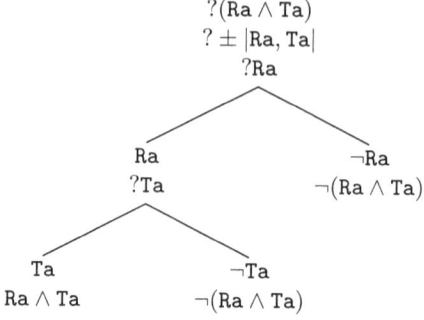

Fig. 11.2. *The e-scenario which is embedded into the previous one.*

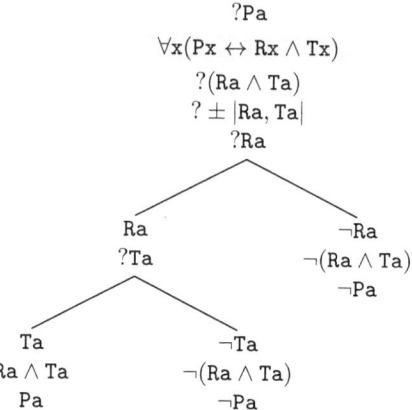

Fig. 11.3. *The result of embedding.*

the solution (given the initial premise) is Pa, and once ¬(Ra ∧ Ta) holds, the solution is ¬Pa (again, given the premise). But question ?(Ra∧Ta) is the principal question of the e-scenario depicted in Figure 11.2. This e-scenario, in turn, shows how the query can be resolved step by step by asking and answering further queries, and that one gets the answer Ra∧Ta if one has got Ra and then Ta, and gets the answer ¬(Ra∧Ta) if one has got either only ¬Ra or first Ra and then ¬Ta. The e-scenario depicted by Figure 11.3 carries the above pragmatic information in addition to that already provided by the e-scenario displayed in Figure 11.1. On the other hand, the e-scenario can be constructed out of the e-scenarios displayed in Figures 11.1 and 11.2 *purely syntactically*.

How? Consider the paths of the e-scenario displayed in Figure 11.1:

$$?Pa, \forall x(Px \leftrightarrow Rx \wedge Tx), ?(Ra \wedge Ta), Ra \wedge Ta, Pa \qquad (11.1)$$

$$?Pa, \forall x(Px \leftrightarrow Rx \wedge Tx), ?(Ra \wedge Ta), \neg(Ra \wedge Ta), \neg Pa \qquad (11.2)$$

Take the query-answer segment of (11.1):

$$?(Ra \wedge Ta), Ra \wedge Ta \qquad (11.3)$$

11.1 Embedding

There is only one path of the e-scenario depicted in Figure 11.2 that leads to Ra ∧ Ta, namely:

$$?(Ra \wedge Ta), ? \pm |Ra, Ta|, ?Ra, Ra, ?Ta, Ta, Ra \wedge Ta \qquad (11.4)$$

Let us replace the query-answer segment (11.3) of (11.1) with the sequence (11.4). The result is (to improve readability, the segment (11.4) of (11.5) is underlined):

$$?Pa, \forall x(Px \leftrightarrow Rx \wedge Tx), \underline{?(Ra \wedge Ta), ? \pm |Ra, Ta|, ?Ra, Ra, ?Ta, Ta, Ra \wedge Ta}, Pa \qquad (11.5)$$

Now take the query-answer segment of (11.2):

$$?(Ra \wedge Ta), \neg(Ra \wedge Ta) \qquad (11.6)$$

There are two paths of the e-scenario displayed in Figure 11.2 which lead to ¬(Ra ∧ Ta):

$$?(Ra \wedge Ta), ? \pm |Ra, Ta|, ?Ra, \neg Ra, \neg(Ra \wedge Ta) \qquad (11.7)$$
$$?(Ra \wedge Ta), ? \pm |Ra, Ta|, ?Ra, Ra, ?Ta, \neg Ta, \neg(Ra \wedge Ta) \qquad (11.8)$$

First, we replace the query-answer segment (11.6) of (11.2) with the path (11.7) of the e-scenario for the query. We get:

$$?Pa, \forall x(Px \leftrightarrow Rx \wedge Tx), \underline{?(Ra \wedge Ta), ? \pm |Ra, Ta|, ?Ra, \neg Ra, \neg(Ra \wedge Ta)}, \neg Pa \qquad (11.9)$$

Second, we replace the segment (11.6) of (11.2) with the path (11.8). The result is:

$$?Pa, \forall x(Px \leftrightarrow Rx \wedge Tx), \underline{?(Ra \wedge Ta), ? \pm |Ra, Ta|, ?Ra, Ra, ?Ta, \neg Ta,}$$
$$\underline{\neg(Ra \wedge Ta)}, \neg Pa \qquad (11.10)$$

Each of (11.5), (11.9), (11.10) is an e-derivation of a direct answer to question ?Pa on the basis of (the set whose only element is) ∀x(Px ↔ Rx ∧ Tx). They, jointly, constitute the e-scenario displayed in Figure 11.3.

The embedded scenario was a pure one, that is, the set of its initial premises was empty. A slight complication arises when an embedded e-scenario does not have this property. Let us illustrate this with an example.

Example 11.2. Take the e-scenario displayed in Figure 11.4 (already presented in section 9.5 of Chapter 9). Let us embed the e-scenario depicted in Figure 11.5 into the e-scenario displayed in Figure 11.4; the embedding takes place with respect to the query ?t. The result of embedding is displayed in Figure 11.6.

Underlining indicates the items of the e-scenario which has been embedded that occur in the resultant e-scenario; we used underlining in order to improve readability only.

140 11 Operations on E-scenarios

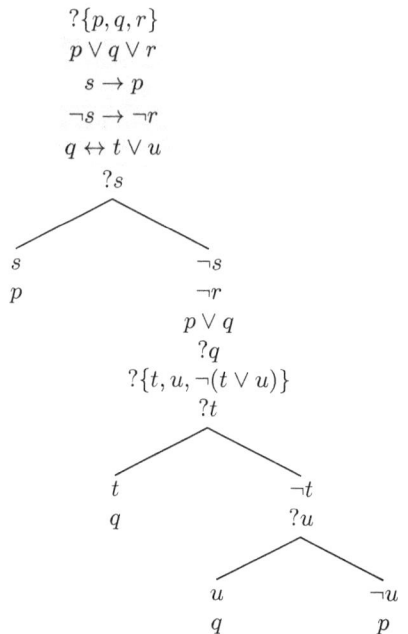

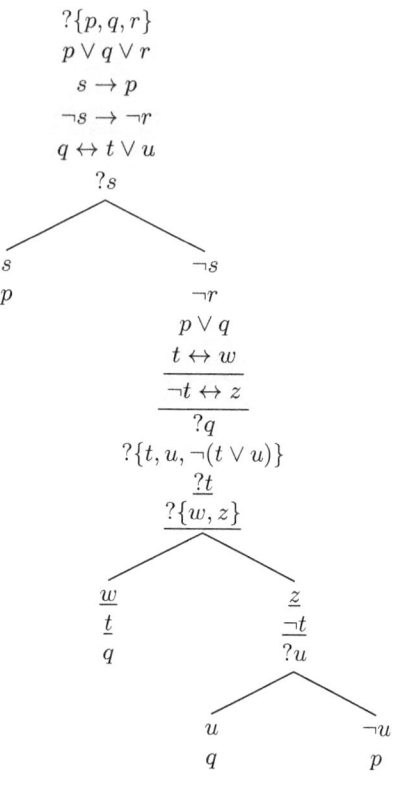

Fig. 11.4. *An e-scenario subjected to embedding.*

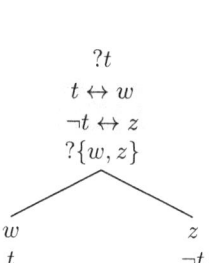

Fig. 11.5. *An e-scenario to be embedded.*

Fig. 11.6. *The result of embedding of the e-scenario depicted in Figure 11.5 into the e-scenario displayed in Figure 11.4.*

The paths of the e-scenario displayed in Figure 11.5:

$$?t, t \leftrightarrow w, \neg t \leftrightarrow z, ?\{w, z\}, w, t \qquad (11.11)$$

$$?t, t \leftrightarrow w, \neg t \leftrightarrow z, ?\{w, z\}, z, \neg t \qquad (11.12)$$

share the segment:

$$t \leftrightarrow w, \neg t \leftrightarrow z \qquad (11.13)$$

whose terms are initial premises that occur in the e-scenario before its first query; let us call it "initial d-segment". This segment occurs in the resultant e-scenario (depicted in Figure 11.6) just before the first auxiliary question that precedes the query $?t$ for which we embedded. Moreover, the following segments of the e-scenario subjected to embedding (displayed in Figure 11.4):

$$?t, t \qquad (11.14)$$

$$?t, \neg t \qquad (11.15)$$

are replaced in the resultant e-scenario by the sequences of wffs:

$$?t, ?\{w, z\}, w, t \qquad (11.16)$$

$$?t, ?\{w, z\}, z, \neg t \qquad (11.17)$$

Both (11.16) and (11.17) begin with the query with respect to which embedding takes place/the principal question of the embedded e-scenario, and then comprise segments of paths (11.11) and (11.12) of the embedded e-scenario which immediately succeed the initial d-segment and agree with the replaced query-answer segments (11.14) and (11.15) as to last terms. Such a move warrants that the resultant sequences of wffs are e-derivations. Recall that, according to clause (1c) of Definition 9.1 of e-derivation, an auxiliary question must be immediately followed either by a direct answer to it or by a question. For this reason we put initial d-segments of embedded e-scenarios just before the first auxiliary question that precedes the query for which we embed, or, if there are no such auxiliary questions, just before the query itself.

Observe that the e-scenarios depicted in Figures 11.4 and 11.6 do not differ which respect to their leftmost paths. These paths, however, do not "go through" the query for which we embed.

11.1.1 A formal account of embedding

In order to give a formal account of embedding we need some auxiliary notions as well as notational conventions.

I. Let Σ be an e-scenario for Q relative to X, and $\mathbf{s} = \mathbf{s}_1, \ldots, \mathbf{s}_n$ be a path of Σ. For each term $\mathbf{s}_m$ of $\mathbf{s}$ we define the set $\Sigma_{[\mathbf{s},\mathbf{s}_m]}$ of paths of Σ that "go through" $\mathbf{s}_m$.

$$\Sigma_{[\mathbf{s},\mathbf{s}_m]} = \{\mathbf{s}^* \in \Sigma : \mathbf{s}_i^* = \mathbf{s}_i \text{ for } i = 1, \ldots, m\} \qquad (11.18)$$

$\Sigma_{[\mathbf{s},\mathbf{s}_m]}$ is thus the set of paths of Σ which have the wff $\mathbf{s}_m$ as m-th term and agree with $\mathbf{s}$ as to previous terms.

We define:
$$\widehat{\Sigma}_{[\mathbf{s},\mathbf{s}_m]} = \Sigma \setminus \Sigma_{[\mathbf{s},\mathbf{s}_m]} \tag{11.19}$$

$\widehat{\Sigma}_{[\mathbf{s},\mathbf{s}_m]}$ is the set of paths of Σ that do not "go through" $\mathbf{s}_m$. Clearly we have:
$$\Sigma_{[\mathbf{s},\mathbf{s}_m]} \cup \widehat{\Sigma}_{[\mathbf{s},\mathbf{s}_m]} = \Sigma \tag{11.20}$$

II. Let $\mathbf{s} = \mathbf{s}_1, \ldots, \mathbf{s}_n$ be a path of Σ, and $\mathbf{s}_k$ be a query of $\mathbf{s}$. Path $\mathbf{s}$ can be displayed/analysed as:
$$\gamma_{[\mathbf{s}]} \,{}' \mathbf{s}_j \,{}' \epsilon_{[\mathbf{s}]} \,{}' \mathbf{s}_k, \mathbf{s}_{k+1} \,{}' \zeta_{[\mathbf{s}]} \tag{11.21}$$
where $'$ stands for the concatenation sign and:

1. j is the greatest index lower than k such that $\mathbf{s}_j$ is not an auxiliary question and $\mathbf{s}_{j+1}, \ldots, \mathbf{s}_k$ is a sequence of questions;
2. $\mathbf{s}_{k+1}$ is the direct answer to $\mathbf{s}_k$, that is, the next term of $\mathbf{s}$ after $\mathbf{s}_k$;
3. $\gamma_{[\mathbf{s}]} = \mathbf{s}_1, \ldots, \mathbf{s}_{j-1}$;
4. $\epsilon_{[\mathbf{s}]} = \mathbf{s}_{j+1}, \ldots, \mathbf{s}_{k-1}$;
5. $\zeta_{[\mathbf{s}]} = \mathbf{s}_{k+2}, \ldots, \mathbf{s}_n$.

As for the first condition, observe that j always exists, is unique, and can be calculated as follows. A query of a path is immediately preceded on the path by: (a) the principal question, or (b) a sequence of auxiliary questions that are not queries, possibly a one-term sequence, or (c) a sequence of d-wffs, again possibly a one-term sequence. In the case of (c) we have $j = k - 1$. If (a) holds, we get $j = 1$, hence $\gamma_{[\mathbf{s}]}$ is empty and $\mathbf{s}_1$ is the principal question. If (b) takes place, $j = k - (d+1)$, where d is the number of terms of the sequence of auxiliary questions that immediately precedes $\mathbf{s}_k$. Of course, $\epsilon_{[\mathbf{s}]}$ is empty if $j = k - 1$, and $\zeta_{[\mathbf{s}]}$ is empty if $n = k + 1$. Note that if $\gamma_{[\mathbf{s}]}$ is non-empty, it may involve queries; $\mathbf{s}_k$ is not supposed to be the first query of a path.

Each path $\mathbf{s}^* = \mathbf{s}_1^*, \ldots, \mathbf{s}_u^*$ in $\Sigma_{[\mathbf{s},\mathbf{s}_k]}$ ($\mathbf{s}$ included!) can be displayed/analysed in this manner. Moreover, j is fixed in $\Sigma_{[\mathbf{s},\mathbf{s}_k]}$, since for any $\mathbf{s}, \mathbf{s}^*$ in $\Sigma_{[\mathbf{s},\mathbf{s}_k]}$ we have $\mathbf{s}_i = \mathbf{s}_i^*$ for $i = 1, \ldots, k$.

III. Let Δ be a *complete* e-scenario for question $\mathbf{s}_k$ relative to a set of d-wffs Y. Each path $\mathbf{g} = \mathbf{g}_1, \ldots, \mathbf{g}_w$ of Δ can be displayed/analysed as:
$$\mathbf{s}_k \,{}' \rho_{[\mathbf{g}]} \,{}' \mathbf{g}_h \,{}' \delta_{[\mathbf{g}]} \,{}' B \tag{11.22}$$
where $'$ is the concatenation sign and:

6. $\mathbf{g}_1 = \mathbf{s}_k$;
7. h is the index of the first auxiliary question of $\mathbf{g}$;
8. $\rho_{[\mathbf{g}]} = \mathbf{g}_2, \ldots, \mathbf{g}_{h-1}$;
9. $\delta_{[\mathbf{g}]} = \mathbf{g}_{h+1}, \ldots, \mathbf{g}_{w-1}$;
10. $\mathbf{g}_w = B$.

Thus B is the direct answer to question $\mathbf{s}_k$ that is the endpoint of path $\mathbf{g}$. Again, if $h = 2$, then $\rho_{[\mathbf{g}]}$ is, obviously, empty, and $\delta_{[\mathbf{g}]}$ is empty if $w = h + 1$.

By Corollary 9.8, each path of Δ involves at least one query, and by Corollary 9.9 there exists the first query of Δ which is common to all the paths.

11.1 Embedding   143

Thus h always exists and is either the index (at $\mathbf{g}$) of the first query of Δ or is the least index (again, at $\mathbf{g}$) of an auxiliary question of $\mathbf{g}$ that precedes the first query. But if $\mathbf{g}$ and $\mathbf{g}^*$ are any paths of Δ, then, again by Corollary 9.9, $\rho_{[\mathbf{g}]} = \rho_{[\mathbf{g}^*]}$ and $\mathbf{g}_h = \mathbf{g}_h^*$. Hence h is fixed for Δ as well. Moreover, if we define:

$$\text{ids}_\Delta = \rho_{[\mathbf{g}]} \tag{11.23}$$

$$\text{faq}_\Delta = \mathbf{g}_h \tag{11.24}$$

("ids" alludes to "initial declarative segment", and "faq" to "first auxiliary question"), any path $\mathbf{g}^* = \mathbf{g}_1^*, \ldots, \mathbf{g}_z^*$ of Δ ($\mathbf{g}$ included!) can be displayed/analysed as:

$$\mathbf{s}_k \,'\, \text{ids}_\Delta \,'\, \text{faq}_\Delta \,'\, \delta_{[\mathbf{g}^*]} \,'\, C \tag{11.25}$$

where C is a direct answer to question $\mathbf{s}_k$.

Now let us define:

$$\Delta_B = \{\mathbf{g} \in \Delta : \mathbf{g} = \mathbf{s}_k \,'\, \text{ids}_\Delta \,'\, \text{faq}_\Delta \,'\, \delta_{[\mathbf{g}]} \,'\, B\} \tag{11.26}$$

Δ_B is thus the set of paths of Δ that end with the answer B to the principal question $\mathbf{s}_k$ of Δ. Note that Δ_B need not be a singleton set.

IV. As above, $\mathbf{s}_k$ is assumed to be a query of a path $\mathbf{s}$ of Σ. Consider the following set:

$$\Sigma_{[\mathbf{s},\mathbf{s}_k,B]} = \{\mathbf{s}^* \in \Sigma_{[\mathbf{s},\mathbf{s}_k]} : \mathbf{s}_{k+1}^* = B\} \tag{11.27}$$

where $B \in \text{ds}_k$. Thus, $\Sigma_{[\mathbf{s},\mathbf{s}_k,B]}$ is the set of paths of Σ that "agree" with $\mathbf{s}$ up to their k-th terms and then have the answer B to query $\mathbf{s}_k$ as their $(k+1)$st terms. (The path $\mathbf{s}$ may belong to $\Sigma_{[\mathbf{s},\mathbf{s}_k,B]}$ or not, depending on whether $\mathbf{s}_{k+1} = B$.)

Take an arbitrary but fixed element $\mathbf{t} = \mathbf{t}_1, \ldots, \mathbf{t}_u$ of $\Sigma_{[\mathbf{s},\mathbf{s}_k,B]}$. Since $\Sigma_{[\sigma,\sigma_k,B]}$ is included in $\Sigma_{[\mathbf{s},\mathbf{s}_k]}$, the sequence $\mathbf{t}$ can be displayed/analysed as:

$$\gamma_{[\mathbf{t}]} \,'\, \mathbf{t}_j \,'\, \epsilon_{[\mathbf{t}]} \,'\, \mathbf{t}_k, B \,'\, \zeta_{[\mathbf{t}]} \tag{11.28}$$

Of course, we have:

$$\gamma_{[\mathbf{t}]} \,'\, \mathbf{t}_j \,'\, \epsilon_{[\mathbf{t}]} \,'\, \mathbf{t}_k = \gamma_{[\mathbf{s}]} \,'\, \mathbf{s}_j \,'\, \epsilon_{[\mathbf{s}]} \,'\, \mathbf{s}_k \tag{11.29}$$

and thus question $\mathbf{s}_k$ is also a query of $\mathbf{t}$ preceded in $\mathbf{t}$ by the same sequence of wffs which precedes it in $\mathbf{s}$. In other words, the segments of $\mathbf{s}$ and $\mathbf{t}$ ending with $\mathbf{s}_k$ and $\mathbf{t}_k$, respectively, are identical.

V. Let us now define a certain operation on $\Sigma_{[\mathbf{s},\mathbf{s}_k,B]}$ and Δ_B, symbolized by ⊛:

Definition 11.3. *Let* $\mathbf{t} \in \Sigma_{[\mathbf{s},\mathbf{s}_k,B]}$ *and* $\mathbf{g} \in \Delta_B$.

$$\mathbf{t} \circledast \mathbf{g} = \gamma_{[\mathbf{t}]} \,'\, \mathbf{t}_j \,'\, \text{ids}_\Delta \,'\, \epsilon_{[\mathbf{t}]} \,'\, \mathbf{t}_k \,'\, \text{faq}_\Delta \,'\, \delta_{[\mathbf{g}]} \,'\, B \,'\, \zeta_{[\mathbf{t}]}$$

Since (11.29) is true, we get:

$$\mathbf{t} \circledast \mathbf{g} = \gamma_{[\mathbf{s}]} \,'\, \mathbf{s}_j \,'\, \text{ids}_\Delta \,'\, \epsilon_{[\mathbf{s}]} \,'\, \mathbf{s}_k \,'\, \text{faq}_\Delta \,'\, \delta_{[\mathbf{g}]} \,'\, B \,'\, \zeta_{[\mathbf{t}]} \tag{11.30}$$

for any $\mathbf{t} \in \Sigma_{[\mathbf{s},\mathbf{s}_k,B]}$ and $\mathbf{g} \in \Delta_B$.[1] Recall that the following

---

[1] If $\mathbf{s} \in \Sigma_{[\mathbf{s},\mathbf{s}_k,B]}$, then $\mathbf{s} \circledast \mathbf{g} = \gamma_{[\mathbf{s}]} \,'\, \mathbf{s}_j \,'\, \text{ids}_\Delta \,'\, \epsilon_{[\mathbf{s}]} \,'\, \mathbf{s}_k \,'\, \text{faq}_\Delta \,'\, \delta_{[\mathbf{g}]} \,'\, B \,'\, \zeta_{[\mathbf{s}]}$.

$$\mathbf{g} = \mathbf{s}_k \,'\, \mathtt{ids}_\Delta \,'\, \mathtt{faq}_\Delta \,'\, \delta_{[\mathbf{g}]} \,'\, B$$

Thus ⊛ puts the initial declarative segment, $\mathtt{ids}_\Delta$, of $\mathbf{g}$ immediately before the first auxiliary question that precedes[2] the query $\mathbf{s}_k$ or, if there is no such auxiliary question, just before $\mathbf{s}_k$ itself, and then replaces the query-answer segment $\mathbf{s}_k, B$ with the sequence:

$$\mathbf{s}_k \,'\, \mathtt{faq}_\Delta \,'\, \delta_{[\mathbf{g}]} \,'\, B$$

We have already used this kind of transformation in our analysis of Example 11.2. When $\mathtt{ids}_\Delta$ is empty, the segment of $\mathbf{t}$ that precedes the query for which we embed remains unchanged. This special case of ⊛ had been used in Example 11.1.

One may ask why the segment $\mathtt{faq}_\Delta$ of $\mathbf{g}$ is required to be "moved" separately, that is, to be situated just before the first auxiliary question that precedes question/query $\mathbf{s}_k$ or, if there is no such auxiliary question, immediately before question $\mathbf{s}_k$ itself. The answer is: this warrants that $\mathbf{t} \circledast \mathbf{g}$ is an e-derivation. Due to clause (1c) of Definition 9.1 of e-derivation, an auxiliary question must be followed either by a direct answer or by a question. Question $\mathbf{s}_k$ is an auxiliary question of $\mathbf{t}$ and placing the segment $\mathtt{faq}_\Delta$ immediately after $\mathbf{s}_k$ (as it has occurred in the e-scenario which is embedded) would result in violating clause (1c) of Definition 9.1.

VI. We also need the following:

Definition 11.4. *Let $\mathbf{s}$ be a path of an e-scenario for question Q, $\mathbf{s}_k$ be a query of $\mathbf{s}$, and Δ be a complete e-scenario for question $\mathbf{s}_k$.*

1. $\Sigma_{[\mathbf{s},\mathbf{s}_k]}^{\Delta_B, \circledast} = \{\mathbf{t} \circledast \mathbf{g} : \mathbf{t} \in \Sigma_{[\mathbf{s},\mathbf{s}_k, B]} \text{ and } \mathbf{g} \in \Delta_B\}$
2. $\Sigma_{[\mathbf{s},\mathbf{s}_k]}^{\Delta, \circledast} = \bigcup_{B \in \mathtt{ds}_k} \Sigma_{[\mathbf{s},\mathbf{s}_k]}^{B, \circledast}$

Observe that $\Sigma_{[\mathbf{s},\mathbf{s}_k]}^{\Delta_B, \circledast}$ is a singleton set only if both $\Sigma_{[\mathbf{s},\mathbf{s}_k, B]}$ and Δ_B are singleton sets[3], otherwise $\Sigma_{[\mathbf{s},\mathbf{s}_k]}^{\Delta_B, \circledast}$ has more than one element. However, $\Sigma_{[\mathbf{s},\mathbf{s}_k]}^{B, \circledast}$ is always a finite set, since both Σ and Δ are, as any e-scenarios, finite sets. Since Δ is an e-scenario for a query, and, by Corollary 9.7, each query is a question with a finite number of direct answers, $\Sigma_{[\mathbf{s},\mathbf{s}_k]}^{\Delta, \circledast}$ is always finite. On the other hand, each question is assumed to have at least two direct answers and thus $\Sigma_{[\mathbf{s},\mathbf{s}_k]}^{\Delta, \circledast}$ is never a singleton set.

VII. By applying the concepts introduced above we define embedding by:

Definition 11.5 (Embedding). *Let $\mathbf{s}_k$ be a query of a path $\mathbf{s}$ of an e-scenario Σ, and Δ be a complete e-scenario for question $\mathbf{s}_k$.*

$$\mathrm{EMB}(\Delta/\mathbf{s}, \mathbf{s}_k, \Sigma) = \widehat{\Sigma}_{[\mathbf{s},\mathbf{s}_k]} \cup \Sigma_{[\mathbf{s},\mathbf{s}_k]}^{\Delta, \circledast}$$

---

[2] In $\mathbf{t}$ and in $\mathbf{s}$; recall that (11.29) holds and $\mathbf{s}_k = \mathbf{t}_k$ is the case.
[3] The former happens iff question $\mathbf{s}_k$ is the "last" query of $\mathbf{t}$, the latter iff there is only one path of Δ that leads to B. Neither of those situations is a rule.

The inscription "EMB(Δ/s, s_k, Σ)" reads "the result of embedding Δ into Σ with respect to query s_k of path s of Σ".

Let us stress that embedding is defined only for the case Δ is a complete e-scenario for the query. Note also that embedding is akin to replacement rather than to substitution: a given question can occur as a query on many different paths or occur many times at the same path, but embedding always pertains to a fixed occurrence of the question *as a query* at a given path, and affects a cluster of paths. To speak generally, embedding is not "global", but is always "local".

Recall that $\widehat{\Sigma}_{[s,s_k]}$ is the set of paths of Σ that do not "go through" the query s_k. These paths, if there are any, remain unaffected by embedding.[4]

Clearly we have:

Corollary 11.6. EMB(Δ/s, s_k, Σ) = EMB(Δ/t, t_k, Σ), *for any* $t \in \widehat{\Sigma}_{[s,s_k]}$.

A thorough examination of the constructions presented above leads us to:

Theorem 11.7 (*Embedding Theorem*). *Let Σ be an e-scenario for a question Q relative to a set of d-wffs X, and let s_k be a query of a path s of Σ. Let Δ be a complete e-scenario for question s_k relative to a set of d-wffs Y. EMB(Δ/s, s_k, Σ) is an e-scenario for Q relative to $X \cup Y$ if the following conditions hold:*

1. $Y \cap dQ = \emptyset$, *and*
2. *for each question Q^* of Δ :* $dQ^* \neq dQ$.

Conditions (1) and (2) are indispensable, due to clause (4) of Definition 9.4 of e-scenarios, and clause (1a) of Definition 9.1 of e-derivations. Δ has to be complete because otherwise clause (3a) of Definition 9.4 would not hold.

11.1.2 A procedural account of embedding

Embedding has been defined with all the details since this enables us to prove some feasibility as well as reducibility results (see the next chapter). However, to embed an e-scenario into an e-scenario is, in practice, an easy enterprise. Below we sketch a simple procedure which can be turned into an algorithm.

GOAL: Σ is an e-scenario for Q relative to X. The task is to perform embedding in Σ w.r.t. query s_k of path s of Σ. What is embedded is a complete e-scenario Δ for the question/query s_k.

PREPARATORY STEPS:

1. We identify the initial declarative segment ids_Δ of Δ and the first auxiliary question faq_Δ of Δ.
2. We establish some order, $B_1, \ldots, B_n$, of the elements of ds_k, i.e. direct answers to the question s_k.

---

[4] $\widehat{\Sigma}_{[s,s_k]}$ is empty if s_k is the first query of Σ. If we embed with respect to the first query, all the paths are affected.

3. For each direct answer B_i to $\mathbf{s}_k$ we identify Δ_{B_i}, i.e. the set comprising all the paths of Δ that end with B_i. Then, for each Δ_{B_i}, we build the set $\Delta^\star_{B_i}$ of segments of elements of Δ_{B_i} that begin with $\mathtt{faq}_\Delta$ and end with B_i.
4. For each $i = 1, \ldots, n$, we establish some order of elements of $\Delta^\star_{B_i}$.
5. For each set of sequences $\Sigma_{[\mathbf{s},\mathbf{s}_k,B_i]}$ we establish some order of the elements of the set.

FIRST STEP: We replace each sequence $\mathbf{t}$ in $\Sigma_{[\mathbf{s},\mathbf{s}_k,B_i]}$ (i.e. each path of Σ "going through" the query $\mathbf{s}_k$ and its answer B_i) with the sequence which differs from $\mathbf{t}$ only in having the first (in the order previously established) element of $\Delta^\star_{B_i}$ at the place at which the query/answer segment $\mathbf{s}_k, B_i$ occurs in $\mathbf{t}$.

We proceed analogously with respect to the consecutive elements of $\Delta^\star_{B_i}$ (if there are any). We stop when all the elements have been already used w.r.t. all the paths in $\Sigma_{[\mathbf{s},\mathbf{s}_k,B_i]}$.

Having the procedure completed for paths in $\Sigma_{[\mathbf{s},\mathbf{s}_k,B_i]}$, we proceed in an analogous manner with sequences in $\Sigma_{[\mathbf{s},\mathbf{s}_k,B_{i+1}]}$ w.r.t. $\Delta^\star_{B_{i+1}}$.

We start with $i = 1$, for each $1 \leq i \leq n$ proceed according to the order of sequences in $\Sigma_{[\mathbf{s},\mathbf{s}_k,B_i]}$, and stop when all the required operations on sequences in $\Sigma_{[\mathbf{s},\mathbf{s}_k,B_n]}$ have been performed.

SECOND STEP: We check if $\mathbf{s}_k$ is immediately preceded on $\mathbf{s}$ by the principal question Q.

1. If yes, we put $\mathtt{ids}_\Delta$ just after Q.
2. If not, we check if $\mathbf{s}_k$ is immediately preceded on $\mathbf{s}$ by a non-empty sequence of auxiliary questions.
   a. If yes, we put $\mathtt{ids}_\Delta$ just before the first term of the sequence.
   b. If not, we put $\mathtt{ids}_\Delta$ just before $\mathbf{s}_k$.

The above operation is performed upon each sequence obtained in the first step.

The outcome is an e-scenario for Q relative to the union of X and the background of the embedded e-scenario Δ given that the conditions specified by Theorem 11.7 are met.

11.2 Contraction

The idea which underlies the concept of contraction is the following. We have an e-scenario Σ for a question Q relative to a set of d-wffs X, a path $\mathbf{s} = \mathbf{s}_1, \ldots, \mathbf{s}_n$ of Σ, and a query $\mathbf{s}_k$ of $\mathbf{s}$. The d-wff $\mathbf{s}_{k+1}$ which is the $k+1$-st term of $\mathbf{s}$ is thus a direct answer to question/query $\mathbf{s}_k$. We assume that $\mathbf{s}_k$ has been answered with $\mathbf{s}_{k+1}$. This cancels the query $\mathbf{s}_k$ and makes the d-wff $\mathbf{s}_{k+1}$ a new initial premise. Σ contracts with respect to "new" information carried by $\mathbf{s}_{k+1}$: the paths of Σ which go through the other answers to query $\mathbf{s}_k$ become irrelevant and thus are deleted, while the paths which go through $\mathbf{s}_k$ *and* $\mathbf{s}_{k+1}$ transform accordingly.

We first define contraction in general terms and then illustrate the definition with examples.

11.2.1 A formal account of contraction

I. Let **s** be a path of an e-scenario Σ for Q relative to X, and let $\mathbf{s}_k$ be a query of **s**. Clearly, **s** belongs to $\Sigma_{[\mathbf{s},\mathbf{s}_{k+1}]}$, but usually $\Sigma_{[\mathbf{s},\mathbf{s}_{k+1}]}$ involves also some other path(s) of Σ.[5]

Let $\mathbf{t} = \mathbf{t}_1, \ldots, \mathbf{t}_u$ be an arbitrary but fixed element of $\Sigma_{[\mathbf{s},\mathbf{s}_{k+1}]}$. According to what has been said in section 11.1.1, path **t** can be displayed/analysed as:[6]

$$\gamma_{[\mathbf{t}]} \,'\, \mathbf{t}_j \,'\, \epsilon_{[\mathbf{t}]} \,'\, \mathbf{t}_k, \mathbf{t}_{k+1} \,'\, \zeta_{[\mathbf{t}]} \tag{11.31}$$

where j is the greatest index lower than k such that $\mathbf{t}_j$ is not an auxiliary question, and $\epsilon_{[\mathbf{t}]}$ is a (possibly empty) sequence of auxiliary questions that are not queries. Of course, we have:

$$\gamma_{[\mathbf{t}]} \,'\, \mathbf{t}_j \,'\, \epsilon_{[\mathbf{t}]} \,'\, \mathbf{t}_k, \mathbf{t}_{k+1} = \gamma_{[\mathbf{s}]} \,'\, \mathbf{s}_j \,'\, \epsilon_{[\mathbf{s}]} \,'\, \mathbf{s}_k, \mathbf{s}_{k+1} \tag{11.32}$$

As for $\zeta_{[\mathbf{t}]}$, there are three possibilities:

(a) $u = k+1$ and thus $\zeta_{[\mathbf{t}]}$ is empty;
(b) $u > k+1$ and no term of $\zeta_{[\mathbf{t}]}$ is a question;
(c) $u > k+1$ and at least one term of $\zeta_{[\mathbf{t}]}$ is a question.

If (a) or (b) hold, $\mathbf{t}_k$ is the last query of **t**. In the case of (c) the sequence $\zeta_{[\mathbf{t}]}$ can be displayed/analysed as:

$$\check{\zeta}_{[\mathbf{t}]} \,'\, \hat{\zeta}_{[\mathbf{t}]} \tag{11.33}$$

where $\check{\zeta}_{[\mathbf{t}]}$ is a (possibly empty) sequence of d-wffs and $\hat{\zeta}_{[\mathbf{t}]}$ is a sequence of wffs whose first term is a question.

II. We define a certain operation $\ominus$ on $\Sigma_{[\mathbf{s},\mathbf{s}_{k+1}]}$:

Definition 11.8. *Let* $\mathbf{t} \in \Sigma_{[\mathbf{s},\mathbf{s}_{k+1}]}$.

1. If $u = k+1$, then:

$$\ominus \mathbf{t} = \gamma_{[\mathbf{t}]} \,'\, \mathbf{t}_j \,'\, \mathbf{t}_{k+1}$$

2. If $u > k+1$ and no term of $\zeta_{[\mathbf{t}]}$ is a question, then:

$$\ominus \mathbf{t} = \gamma_{[\mathbf{t}]} \,'\, \mathbf{t}_j \,'\, \mathbf{t}_{k+1} \,'\, \zeta_{[\mathbf{t}]}$$

3. If $u > k+1$ and at least one term of $\zeta_{[\mathbf{t}]}$ is a question, then:

$$\ominus \mathbf{t} = \gamma_{[\mathbf{t}]} \,'\, \mathbf{t}_j \,'\, \mathbf{t}_{k+1} \,'\, \check{\zeta}_{[\mathbf{t}]} \,'\, \epsilon_{[\mathbf{t}]} \,'\, \hat{\zeta}_{[\mathbf{t}]}$$

Some comments are in order. In each of the above cases the question/query $\mathbf{t}_k$ is deleted. When $\mathbf{t}_k$ is the last query of **t**, the auxiliary questions that immediately precede $\mathbf{t}_k$ in **t**, if there are any, are deleted as well; more formally, the segment $\epsilon_{[\mathbf{t}]}$ is deleted. Otherwise the segment $\check{\zeta}_{[\mathbf{t}]}$ is placed immediately after the answer $\mathbf{t}_{k+1}$ and followed by the segments $\epsilon_{[\mathbf{t}]}$ and $\hat{\zeta}_{[\mathbf{t}]}$ (in this order).

---

[5] $\Sigma_{[\mathbf{s},\mathbf{s}_{k+1}]}$ is a singleton set only if $\mathbf{s}_k$ is the last query of **s**.
[6] Again, because $\Sigma_{[\mathbf{s},\mathbf{s}_{k+1}]}$ is included in $\Sigma_{[\mathbf{s},\mathbf{s}_k]}$.

11 Operations on E-scenarios

Such transformation secure $\ominus \mathbf{t}$ to be an e-derivation.[7] Needless to say, $\ominus \mathbf{t}$ obtained according to (1) or (2) above is an e-derivation as well.

III. Next, we introduce:

Definition 11.9. *Let* $\mathbf{s}$ *be a path of an e-scenario* Σ, *and* $\mathbf{s}_k$ *be a query of* $\mathbf{s}$.

$$\Sigma^{\ominus}_{[\mathbf{s},\mathbf{s}_{k+1}]} = \{\ominus \mathbf{t} : \mathbf{t} \in \Sigma_{[\mathbf{s},\mathbf{s}_{k+1}]}\}$$

Contraction can now be defined by:

Definition 11.10 (Contraction). *Let* $\mathbf{s}_k$ *be a query of a path* $\mathbf{s}$ *of an e-scenario* Σ, *and* $\mathbf{s}_{k+1}$ *be the direct answer to* $\mathbf{s}_k$ *occurring on* $\mathbf{s}$.

$$\mathrm{CTR}(\mathbf{s}_{k+1} \parallel \mathbf{s}, \mathbf{s}_k, \Sigma) = \widehat{\Sigma}_{[\mathbf{s},\mathbf{s}_k]} \cup \Sigma^{\ominus}_{[\mathbf{s},\mathbf{s}_{k+1}]}$$

The inscription "$\mathrm{CTR}(\mathbf{s}_{k+1} \parallel \mathbf{s}, \mathbf{s}_k, \Sigma)$" reads "the result of contracting Σ by the answer $\mathbf{s}_{k+1}$ to query $\mathbf{s}_k$ of path $\mathbf{s}$ of Σ."

Recall that $\widehat{\Sigma}_{[\mathbf{s},\mathbf{s}_k]}$ is the set of paths of Σ that do not "go through" the query $\mathbf{s}_k$. Similarly as in the case of embedding, the paths belonging to $\widehat{\Sigma}_{[\mathbf{s},\mathbf{s}_k]}$, if there are any, remain unaffected.

Observe that the following holds:

Corollary 11.11. *If* $\mathbf{t}^* \in \Sigma_{[\mathbf{s},\mathbf{s}_k]}$ *is a path of* Σ *such that* $\mathbf{t}^*_{k+1} \neq \mathbf{s}_{k+1}$, *then* $\mathbf{t}^* \notin \Sigma^{\ominus}_{[\mathbf{s},\mathbf{s}_{k+1}]}$.

Thus no path of Σ that goes through a (direct) answer to the query $\mathbf{s}_k$ other than $\mathbf{s}_{k+1}$ belongs to $\mathrm{CTR}(\mathbf{s}_{k+1} \parallel \mathbf{s}, \mathbf{s}_k, \Sigma)$: these paths are deleted. This is how it should be.

We also have:

Corollary 11.12. $\mathrm{CTR}(\mathbf{s}_{k+1} \parallel \mathbf{s}, \mathbf{s}_k, \Sigma) = \mathrm{CTR}(\mathbf{t}_{k+1} \parallel \mathbf{t}, \mathbf{t}_k, \Sigma)$, *for any* $\mathbf{t} \in \Sigma_{[\mathbf{s},\mathbf{s}_{k+1}]}$.

IV. When an e-scenario is contracted, the result need not be an e-scenario. However, the following holds:

Theorem 11.13 (Contraction Theorem). *Let* Σ *be an e-scenario for a question* Q *relative to a set of d-wffs* X, *and let* $\mathbf{s}_k$ *be a query of a path* $\mathbf{s}$ *of* Σ. $\mathrm{CTR}(\mathbf{s}_{k+1} \parallel \mathbf{s}, \mathbf{s}_k, \Sigma)$ *is an e-scenario for* Q *relative to* $X \cup \{\mathbf{s}_{k+1}\}$ *if*

1. $\mathbf{s}_{k+1} \notin \mathrm{d}Q$ *and*
2. $\widehat{\Sigma}_{[\mathbf{s},\mathbf{s}_k]} \neq \emptyset$ *or* $\Sigma_{[\mathbf{s},\mathbf{s}_k,\mathbf{s}_{k+1}]}$ *involves at least two queries.*

When question $\mathbf{s}_k$ is the only query of Σ, $\mathrm{CTR}(\mathbf{s}_{k+1} \parallel \mathbf{s}, \mathbf{s}_k, \Sigma)$ is a singleton set whose only element is an e-derivation of a direct answer to Q on the basis of the set $X \cup \{\mathbf{s}_{k+1}\}$. The clause (2) is equivalent to: "$\mathrm{CTR}(\mathbf{s}_{k+1} \parallel \mathbf{s}, \mathbf{s}_k, \Sigma)$ has at least two elements".

---

[7] It is possible that a question of $\hat{\zeta}_{[\mathbf{t}]}$ occurs in $\mathbf{t}$ because it is implied by a question of $\epsilon_{[\mathbf{t}]}$ on the basis of a set of d-wffs which includes element(s) of $\check{\zeta}_{[\mathbf{t}]}$.

11.2.2 Examples of contraction

Let us now present some examples. We will operate upon the e-scenario depicted in Figure 11.7.[8]

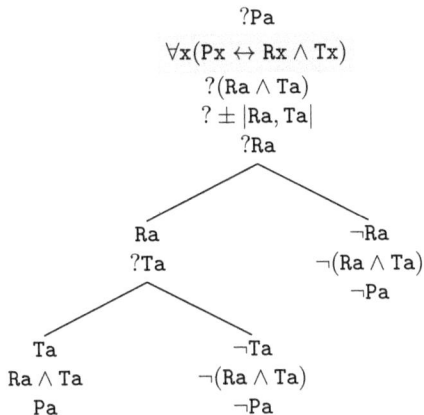

Fig. 11.7. *An example of an e-scenario.*

Example 11.14. Take the e-scenario displayed in Figure 11.7. We contract by the answer Ta to the last query, ?Ta, of the leftmost path. The result is depicted in Figure 11.8. Observe that the path which led through ¬Ta disappears, while the leftmost path transforms accordingly.

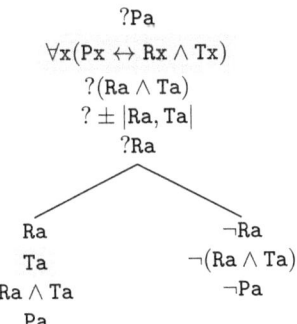

Fig. 11.8. *An e-scenario resulting from the e-scenario displayed in Figure 11.7 by contraction; we contract by the answer Ta to the query ?Ta.*

Example 11.15. Again, we consider the e-scenario displayed in Figure 11.7 and its leftmost path, but this time we take the query ?Ra, and we contract by the answer Ra to it. As a result the rightmost path disappears, question ?Ra is deleted and auxiliary questions that have preceded it now occur after the answer Ra. The outcome is presented in Figure 11.9.

---
[8] Already presented at page 138.

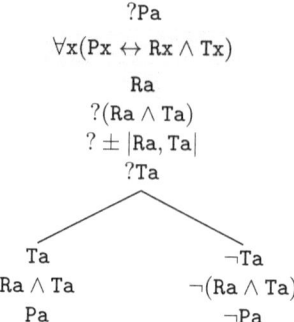

?Pa
∀x(Px ↔ Rx ∧ Tx)
Ra
?(Ra ∧ Ta)
? ± |Ra, Ta|
?Ta

Ta   ¬Ta
Ra ∧ Ta   ¬(Ra ∧ Ta)
Pa   ¬Pa

Fig. 11.9. *An e-scenario resulting from the e-scenario displayed in Figure 11.7 by contraction; we contract by the answer* **Ra** *to the query* ?**Ra**.

Example 11.16. This time we contract the e-scenario displayed in Figure 11.7 with respect to its rightmost path, by the answer ¬Ra to the first query, ?Ra. The result, presented by Figure 11.10, is *not* an e-scenario: it is a singleton set whose only element is an e-derivation of a direct answer to the principal question.

?Pa
∀x(Px ↔ Rx ∧ Tx)
¬Ra
¬(Ra ∧ Ta)
¬Pa

Fig. 11.10. *An example of a result of contraction that is not an e-scenario.*

12
Querying Atomically

Queries of e-scenarios are always questions with finite sets of direct answers. Besides the finiteness requirement, there is no general restriction on the number of direct answers to a query. But if the number is large, the number of paths of an e-scenario is large as well, since, for each direct answer to a query, the e-scenario contains a path or paths which go through the answer, and paths going through distinct answers are distinct. So it is a rational strategy to build e-scenarios whose queries have relatively small sets of direct answers. E-scenarios with queries being *binary questions*, that is, questions having only two direct answers, seem to constitute a privileged class here.

In this chapter we will focus on e-scenarios whose queries are *atomic*, that is, are *atomic yes-no questions*. By an atomic yes-no question we mean a yes-no question whose set of direct answers comprises an atom and its negation. As long as natural languages are concerned, atoms are to be identified with (declarative) sentences in which no propositional connective, modal operator, or quantifier occurs. In the case of propositional languages atoms are just propositional variables. When first-order languages are taken into consideration, atoms are wffs made up of predicates and closed terms.

We will show that, in some cases, e-scenarios which involve non-atomic queries can be transformed, in a systematic manner, into e-scenarios whose queries are atomic. Moreover, relevance, in its broadest sense, is retained: the queries of the "new" scenario are based on atoms which have occurred in the queries of the "old" one.

We are going to stay within the limits of Classical Logic: propositional connectives will be understood classically, and modalities will be ignored. Moreover, we will be considering only atomic yes-no questions whose negative answers are classical (sentential) negations of affirmative answers.[1] The presented results pertain both to the propositional level and to the first-order level.

---

[1] As we pointed out in section 1.3 of Chapter 1, many yes-no questions of a natural language, even based on atoms, permit more sophisticated reading(s). We use the label "simple" in order to indicate that the negative answer results from the affirmative one by preceding it with sentential negation.

12 Querying Atomically

As for the propositional level, we use the language $\mathcal{L}^?_{cpl}$. The language $\mathcal{L}^?_{fom}$ is used as exemplary for the first-order case.

Terminology. For convenience, let us introduce/recall some terminological conventions.

The declarative part of $\mathcal{L}^?_{fom}$ is the language of Monadic First-order Logic with Identity (but without function symbols). By *atoms* of $\mathcal{L}^?_{fom}$ we mean d-wffs of the form:

$$Pc$$

where P is a one-place predicate and c is an individual constant, or of the form:

$$c = c^*$$

where c, c^* are individual constants.

$\mathcal{L}^?_{cpl}$ is the language of CPL enriched with questions. An *atom* of $\mathcal{L}^?_{cpl}$ is a propositional variable.

D-wffs (of $\mathcal{L}^?_{cpl}$ or of $\mathcal{L}^?_{fom}$) in which propositional connectives and/or quantifiers occur are said to be *compound*. A d-wff is called *quantifier-free* if no quantifier occurs in it. The *degree* of a d-wff A (in symbols: $deg(A)$) is the number of occurrences of propositional connectives in A. Note that degrees are independent from the numbers of occurrences of quantifiers.

A *sentence* of $\mathcal{L}^?_{fom}$ is a closed d-wff of the language, i.e. a d-wff with no free variable(s).

It is convenient to introduce a general category of whether-questions.[2] By a *whether-question* we will mean an expression of the form:

$$?\{A_1, \ldots, A_n\} \tag{12.1}$$

where $n > 1$ and $A_1, \ldots, A_n$ are nonequiform (i.e. pairwise syntactically distinct) d-wffs of $\mathcal{L}^?_{cpl}$ when $\mathcal{L}^?_{cpl}$ is considered, and sentences of $\mathcal{L}^?_{fom}$ when $\mathcal{L}^?_{fom}$ is taken into account. All questions of $\mathcal{L}^?_{cpl}$ are whether-questions, whereas $\mathcal{L}^?_{fom}$ includes also which-questions, both existential and general. As in the previous chapters, $A_1, \ldots, A_n$ are the only *direct answers* to a question of the form (12.1).

Due to Corollary 9.7, each query of an e-scenario worded in $\mathcal{L}^?_{cpl}$ or in $\mathcal{L}^?_{fom}$ is a whether-question.

A *simple yes-no question* is a whether-question of the form:

$$?\{A, \neg A\} \tag{12.2}$$

A question of the form (12.2) is abbreviated as:

$$?A \tag{12.3}$$

and is said to be *based on* the d-wff A.

---

[2] In Chapter 2 this label was used only in the context of questions of $\mathcal{L}^?_{fom}$.

An *atomic yes-no question* is a simple yes-no question based on an atom. A query of an e-scenario is called *atomic* if it is an atomic yes-no question. An e-scenario is *atomic* if each query of the scenario is atomic, and *non-atomic* otherwise.

In what follows we will be always assuming that e-scenarios are language-homogeneous: when an e-scenario for a question of a language is considered, all the wffs of the e-scenario belong to the language.

12.1 Atomic e-scenarios for quantifier-free whether-questions

A question is called *quantifier-free* if each direct answer to the question is a quantifier-free d-wff. Clearly, each question of $\mathcal{L}^?_{cpl}$ is quantifier-free for trivial reasons. Some questions of $\mathcal{L}^?_{fom}$ are quantifier-free, but some other are not.

We start with a series of lemmas pertaining to quantifier-free whether-questions.

Lemma 12.1. *Let Q be a quantifier-free simple yes-no question based on a compound d-wff. There exists a pure and complete e-scenario for Q such that each query of the e-scenario is an atomic yes-no question based on an atom that occurs in Q.*

Proof. Let C be a quantifier-free d-wff such that $deg(C) \geq 1$, and let $Q = ?C$.

Assume that $deg(C) = 1$. Now take a look at the schemas of standard e-scenarios for classical connectives depicted in section 10.2.1 of Chapter 10. It is easily visible that when C involves only one occurrence of a connective, the corresponding standard e-scenario for question $?C$ is both pure and complete, and that each query of the e-scenario is an atomic yes-no question based on an atom that occurs in the principal question.

Let C be a d-wff of degree n, where $n > 1$.

Induction hypothesis. If A is a quantifier-free d-wff of degree k, where $1 \leq k < n$, then there exists a pure and complete e-scenario for question $?A$ such that each query of the e-scenario is an atomic yes-no question based on an atom that occurs in $?A$.

Suppose that C is of the form $\neg A$. Clearly $deg(A) < n$ and A is quantifier-free. The standard e-scenario for $?\neg A$, let us designate it by Σ^*, falls under the schema displayed in Figure 12.1.

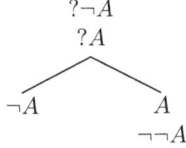

Fig. 12.1. *The case of negation.*

Σ^* is pure and complete. Take the leftmost path, $\mathbf{s}$, of Σ^*. Thus $\mathbf{s}_2 = ?A$. By the induction hypothesis there exists a pure and complete e-scenario Δ for question/query $?A$ whose queries are atomic yes-no questions based on atoms occurring in A. Now let us embed Δ for $\mathbf{s}_2$. To be more precise, consider the following set of sequences of wffs:

$$\text{EMB}(\Delta/\mathbf{s}, \mathbf{s}_2, \Sigma^*) \tag{12.4}$$

Queries of Δ are atomic yes-no questions, whereas question $?\neg A$ is not atomic. Δ is a pure e-scenario. Therefore, by the Embedding Theorem[3], the set (12.4) constitutes an e-scenario for question $?C$. The e-scenario is, obviously, pure and complete. Moreover, each query of (12.4) is an atomic yes-no question based on an atom that occurs in the principal question.

Now suppose that C is of the form $A \wedge B$. The standard e-scenario for $?(A \wedge B)$, let us designate it by Σ°, falls under the schema depicted in Figure 12.2:

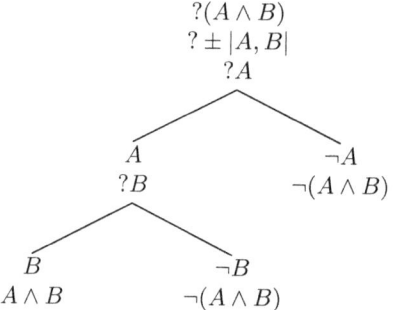

Fig. 12.2. *The case of conjunction.*

Observe that $deg(A) < n$ and $deg(B) < n$. Of course, A and B are quantifier-free.

Assume that neither A nor B is an atom. Thus $0 < deg(A) < n$ and $0 < deg(B) < n$. Hence the induction hypothesis applies to queries of Σ°. Let Δ' and Δ'' be pure and complete e-scenarios for $?A$ and $?B$, respectively; moreover, each query of Δ' is an atomic yes-no question based on an atom that occurs in $?A$, and similarly for Δ'' and $?B$.

Take the leftmost path, $\mathbf{s}$, of Σ°. We have $\mathbf{s}_5 = ?B$. Now let us embed Δ'' for $\mathbf{s}_5$. More precisely, we consider the following set of sequences of wffs:

$$\text{EMB}(\Delta''/\mathbf{s}, \mathbf{s}_5, \Sigma^\circ) \tag{12.5}$$

Both Σ° and Δ'' are pure and complete e-scenarios. Each query of Δ'' is atomic, while Q is not an atomic yes-no question. Thus, by the Embedding Theorem, (12.5) is a pure and complete e-scenario for $?C$. However, $?A$ is (still) not an atomic yes-no question.

---
[3] That is, Theorem 11.7.

12.1 Atomic e-scenarios for quantifier-free whether-questions

In the next step we take the rightmost path, **t**, of the e-scenario (12.5). We have $t_3 = ?A$. We embed Δ' for t_3, that is, we move to:

$$\text{EMB}(\Delta'/\mathbf{t}, t_3, \text{EMB}(\Delta''/\mathbf{s}, s_5, \Sigma^\circ)) \tag{12.6}$$

Since Δ' is pure and complete, all its queries are atomic and $?C$ is not an atomic yes-no question, (12.6) is, by the Embedding Theorem, an e-scenario for $?C$. The construction shows that (12.6) is both pure and complete. Moreover, it shows that each query of (12.6) is an atomic yes-no question based on an atom that occurs in $?C$.

When A is an atom, but B is not, (12.5) constitutes the required e-scenario. When A is a compound d-wff, but B is an atom, the e-scenario sought for is characterized by:

$$\text{EMB}(\Delta'/\mathbf{s}, s_3, \Sigma^\circ) \tag{12.7}$$

The remaining cases are dealt with as follows.

Again, suppose that A and B are compound quantifier-free d-wffs.

If C is of the form $A \to B$, we reason in an analogous manner. The only difference is that we rely upon the schema of the standard e-scenario for implication (cf. Figure 10.4 in Chapter 10).

When C is of the form $A \lor B$, we first take the rightmost path of the standard e-scenario (cf. Figure 10.5) and then the leftmost path of the e-scenario obtained that way.

Finally, if C is of the form $A \leftrightarrow B$, we take the leftmost path of the standard e-scenario (cf. Figure 10.6) and we embed for the second query (i.e. $?B$). Then we take the rightmost path of the e-scenario obtained and we embed for its second query (which is, again, $?B$). Finally, we take the leftmost path of the e-scenario just obtained and we embed for the first query, that is, for $?A$.

When either A or B is an atom, we embed only once. $\square$

Our next lemma is more general than the previous one.

Lemma 12.2. *Let Q be a quantifier-free safe whether-question, but not an atomic yes-no question. There exists a pure and complete e-scenario for Q such that each query of the e-scenario is an atomic yes-no question based on an atom that occurs in Q.*

Proof. Since we have already proved Lemma 12.1, it suffices to consider the case in which Q is a quantifier-free whether-question that is not a simple yes-no question based on a compound d-wff.

By assumption, Q is not an atomic yes-no question. So, in the analysed case, Q is a quantifier-free safe question which is not a simple yes-no question. Let $dQ = \{A_1, \ldots, A_n\}$.

According to what had been said in section 10.2.2 of Chapter 10, there exists a decomposition e-scenario for Q which falls under the schema displayed in Figure 12.3.

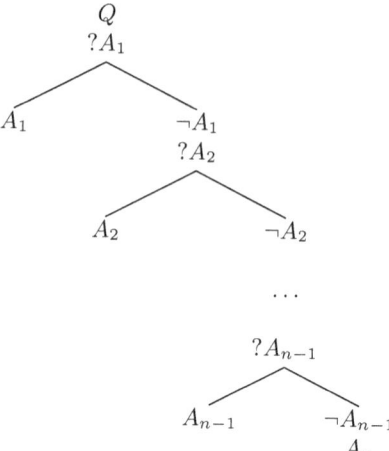

Fig. 12.3. *A schema of the standard decomposition e-scenario for a safe whether-question that is not a simple yes-no question.*

Let us designate the relevant decomposition e-scenario by Σ_0. Observe that Σ_0 is pure and complete, and that all the queries of Σ_0 are simple yes-no questions based on direct answers to Q. Note that the index of i-th query, $?A_i$, of the rightmost path of Σ_0 equals $2i$.

When $A_1, \ldots, A_{n-1}$ are atoms, Σ_0 itself constitutes the atomic e-scenario we are looking for.

Assume that at least one of $A_1, \ldots, A_{n-1}$ is a compound quantifier-free d-wff. The reasoning goes as follows.

We identify all the queries of the rightmost path, $\mathbf{s}$, of Σ_0 which are not atomic yes-no questions. Let they be $?A_{i_1}, \ldots, ?A_{i_m}$, where $i_1 < \ldots < i_m$. Lemma 12.1 warrants that for each question $?A_{i_k}$ ($i_1 \leq i_k \leq i_m$) there exists a corresponding pure and complete e-scenario, say, Δ_{i_k}, whose queries are atomic yes-no questions based on atoms that occur in $?A_{i_k}$. We take the last non-atomic query, $?A_{i_m}$, of the rightmost path of Σ_0, that is, of $\mathbf{s}$. Observe that question $?A_{i_m}$ is the $2i_m$-th term of $\mathbf{s}$. We go to:

$$\text{EMB}(\Delta_{i_m}/\mathbf{s}, \mathbf{s}_{2i_m}, \Sigma_0) \tag{12.8}$$

By the Embedding Theorem, (12.8) constitutes an e-scenario for Q: the conditions (1) and (2) of the theorem are fulfilled, since Q is not a simple yes-no question, while the queries are such questions, and Δ_{i_m} is a pure e-scenario.

If $?A_{i_m}$ is the only non-atomic query of Σ_0, we do nothing, since we have already found an atomic e-scenario with the required properties. Otherwise we repeat the procedure with respect to the e-scenario (12.8). To be more precise, we take the rightmost path, $\mathbf{t}$, of (12.8). Since we have embedded "from the bottom", question $?A_{i_{m-1}}$ is a non-atomic query of $\mathbf{t}$ and, moreover, is $2i_{m-1}$th term of $\mathbf{t}$. We repeat the procedure described above, but this time with respect to $\Delta_{i_{m-1}}$, $\mathbf{t}$, $\mathbf{t}_{2i_{m-1}}$, and (12.8). To be more precise, we go to:

$$\mathrm{EMB}(\Delta_{i_{m-1}}/\mathsf{t}, \mathsf{t}_{2i_{m-1}}, \mathrm{EMB}(\Delta_{i_m}/\mathsf{s}, \mathsf{s}_{2i_m}, \Sigma_0)) \qquad (12.9)$$

It is clear that, after a finite number of steps of the above kind, we will arrive at a pure and complete e-scenario for Q. Moreover, the construction shows that all the queries of the e-scenarios are atomic yes-no questions based on atoms that occur in Q. □

Let us now consider the case of quantifier-free risky whether-questions.

Lemma 12.3. *Let Q be a quantifier-free risky whether-question. There exists a complete e-scenario for Q relative to a disjunction of all the direct answers to Q such that each query of the e-scenario is an atomic yes-no question based on an atom that occurs in Q.*

Proof. Since Q is risky, Q is not a simple yes-no question.

Let $\mathsf{d}Q = \{A_1, \ldots, A_n\}$. Let D be a disjunction of all the elements of $\mathsf{d}Q$. Clearly we have:
$$\{D, \neg A_i\} \models \mathsf{d}Q \setminus \{A_i\}$$
and therefore:
$$\mathbf{Im}(Q, D, ?A_i)$$
for $1 \leq i \leq n-1$. The following holds as well:
$$\{D, \neg A_1, \ldots, \neg A_{n-1}\} \models A_n$$

We construct an e-scenario, $\Sigma^\bullet$, for Q relative to D. The scenario falls under the schema depicted in Figure 12.4.[4] Then we reason similarly as in the proof of Lemma 12.2, but starting with $\Sigma^\bullet$. The second difference stems from the fact that the index of i-th query, $?A_i$, of the rightmost path of $\Sigma^\bullet$ now equals $2i+1$. □

Since each question of $\mathcal{L}^?_{cpl}$ is a quantifier-free whether-question, the results presented by Lemma 12.2 and Lemma 12.3 are pretty general as long as $\mathcal{L}^?_{cpl}$ is concerned. The case of $\mathcal{L}^?_{fom}$ is more complicated. $\mathcal{L}^?_{fom}$ includes also whether-questions which are not quantifier-free as well as which-questions. One cannot generalize the claims of Lemma 12.2 and Lemma 12.3 so that they would pertain to these questions.

12.2 Transforming an e-scenario into an atomic one

Each query of an e-scenario formulated n $\mathcal{L}^?_{cpl}$ or in $\mathcal{L}^?_{fom}$ is a whether-question, but not necessarily a quantifier-free whether-question. In this section we will show that once we have an e-scenario whose queries are quantifier-free questions, but the e-scenario is not already atomic, it can be transformed into an

---
[4] The schema presented by Figure 10.10 (see section 10.2.2 of Chapter 10) is an instance of that depicted in Figure 12.4.

12 Querying Atomically

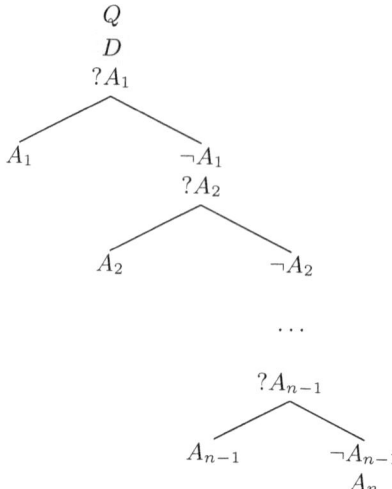

Fig. 12.4. *A schema of a decomposition e-scenario for a risky question.*

atomic one; moreover, the queries of the "new" scenario are atomic yes-no questions based on atoms that occur in the queries of the "old" one. To be more precise, we will prove that, given some conditions are met, such e-scenario exists, and the proof will show how to construct it out of the initial e-scenario.

By a *risky query* we mean a query which is, semantically, a risky question.

Theorem 12.4. *Let Σ be a non-atomic e-scenario for a question Q relative to a set of d-wffs X such that each query of Σ is a quantifier-free question. Let Y be a set whose elements are disjunctions of all the direct answers to risky queries of Σ such that for each risky query of Σ, exactly one disjunction of all the direct answers to the query belongs to Y. If Q is not an atomic yes-no question and $\mathrm{d}Q \cap Y = \emptyset$, then there exists an e-scenario $\Sigma^\triangleleft$ for Q relative to $X \cup Y$ such that all the queries of $\Sigma^\triangleleft$ are atomic yes-no questions based on atoms that occur in the queries of Σ.*

Proof. By assumption, at least one query of Σ is not atomic. Since we consider either language $\mathcal{L}^?_{cpl}$ or language $\mathcal{L}^?_{fom}$, all the queries of Σ are whether-questions which are, by assumption, quantifier-free. So Lemma 12.2 and Lemma 12.3 apply to the case.

Take the rightmost path, $\mathbf{s}$, of Σ, on which a non-atomic query occurs. Consider the last non-atomic query of $\mathbf{s}$, i.e. the non-atomic query of $\mathbf{s}$ whose index in $\mathbf{s}$ is the greatest one. Let k be the index; so the query is $\mathbf{s}_k$. The question $\mathbf{s}_k$ is either safe or risky.

Suppose that $\mathbf{s}_k$ is safe. By Lemma 12.2 there exists a pure and complete e-scenario, Δ_k, for question $\mathbf{s}_k$ whose queries are atomic yes-no questions based on atoms occurring in $\mathbf{s}_k$.

Now suppose that $\mathbf{s}_k$ is risky. Take the disjunction of all the direct answers to $\mathbf{s}_k$ that belongs to Y. Lemma 12.3 warrants that there exists a complete

12.2 Transforming an e-scenario into an atomic one

e-scenario, Δ_k, for question $\mathbf{s}_k$ relative to the disjunction just taken such that the queries of Δ_k are atomic yes-no questions based on atoms that occur in $\mathbf{s}_k$.

Now we go to:
$$\text{EMB}(\Delta_k/\mathbf{s}, \mathbf{s}_k, \Sigma) \tag{12.10}$$

By assumption, Q is not an atomic yes-no question, and $dQ \cap Y = \emptyset$. Thus, by the Embedding Theorem, (12.10) constitutes an e-scenario for Q relative to $X \cup Y$. Observe that each query of a path of (12.10) that goes through $\mathbf{s}_k$ and has an index greater than k (on the path) is an atomic yes-no question based on an atom that occurs in $\mathbf{s}_k$.

If (12.10) already has the required properties, we do nothing. Otherwise we repeat the procedure described above with respect to the rightmost path of (12.10) on which a non-atomic query occurs, and with regard to the last non-atomic query of the path. This gives a new e-scenario for Q relative $X \cup Y$ which either has the required properties or not. In the former case we do nothing, in the latter we repeat the procedure once again. Since the number of non-atomic queries of Σ is finite, after a finite number of steps we arrive at an atomic e-scenario for Q relative to $X \cup Y$. □

What if Σ is an e-scenario for an atomic yes-no question? The construction shows that Σ can be "atomized" on the condition that no direct answer to a query of Σ involves the atom already present in Q. To be more precise, the following holds:

Theorem 12.5. *Let Σ be a non-atomic e-scenario for an atomic yes-no question Q relative to a set of d-wffs X such that: (a) each query of Σ is a quantifier-free question, and (b) the atom occurring in Q does not occur in any query of Σ. Let Y be a set whose elements are disjunctions of all the direct answers to risky queries of Σ such that for each risky query of Σ, exactly one disjunction of all the direct answers to the query belongs to Y. There exists an e-scenario $\Sigma^{\triangleright}$ for Q relative to $X \cup Y$ such that all the queries of $\Sigma^{\triangleright}$ are atomic yes-no questions based on atoms that occur in the queries of Σ.*

Proof. We reason similarly as in the proof of Theorem 12.4. Since Q is atomic, $dQ \cap Y = \emptyset$. Due to clause (b), dQ is not the set of direct answers to any question of an e-scenario which is embedded. So the Embedding Theorem applies in each case. □

Finally, let us note:

Corollary 12.6. *Let Σ be a non-atomic e-scenario for a question Q relative to a set of d-wffs X such that each query of Σ is a quantifier-free safe question. If Q is not an atomic yes-no question, then there exists an e-scenario Σ^* for Q relative to X such that all the queries of Σ^* are atomic yes-no questions based on atoms that occur in the queries of Σ.*

Proof. By Theorem 12.4 (if each query of Σ is safe, then Y is empty). □

Corollary 12.7. *Let Σ be a non-atomic e-scenario for an atomic yes-no question Q relative to a set of d-wffs X such that: (a) each query of Σ is a quantifier-free safe question, and (b) the atom occurring in Q does not occur in any query of Σ. There exists an e-scenario Σ^* for Q relative to X such that all the queries of Σ^* are atomic yes-no questions based on atoms that occur in the queries of Σ.*

Proof. By Theorem 12.5. □

However, the claims of Theorem 12.4 and Theorem 12.5 (as well as of corollaries 12.6 and 12.7) cannot be generalized to these e-scenarios formulated in $\mathcal{L}^?_{fom}$ which involve queries that are not quantifier-free questions. On the other hand, each question of $\mathcal{L}^?_{cpl}$ is a quantifier-free question and thus the results are fully general as long as the propositional case is concerned.

Remarks. The proof of Theorem 12.4 together with the proofs of lemmas 12.1, 12.2 and 12.3 give us some indications concerning a way of transforming a non-atomic e-scenario into an atomic one. The idea is that of *systematic embedding*. We fix Y. At the first step we embed for the last non-atomic query of the rightmost path of Σ on which a non-atomic query occurs. Depending on the form of the query, we build the atomic e-scenario for the query in a way presented in the proof of the appropriate lemma (i.e. Lemma 12.1, or Lemma 12.2, or Lemma 12.3; in the latter case we make use of the relevant element of Y). The outcome is then embedded into Σ. At a consecutive step we embed for the last non-atomic query of the rightmost path of the e-scenario obtained at the previous step on which a non-atomic query occurs. Again, we embed the atomic e-scenario obtained in the way presented in the proof of the appropriate lemma. Moves of this kind are performed until an atomic e-scenario is arrived at. The above schema can be turned into a procedure of transforming a non-atomic e-scenario into an atomic one[5], and then into an algorithm.

The case of the proof of Theorem 12.5 is similar.

Procedures sketched above operate from bottom to top: if a non-atomic query Q^* occurs on a path after a non-atomic query Q, we embed for Q^* first. Yet, one can also perform systematic embedding from top to bottom. The general idea is this. At each step we act upon the leftmost path on which a non-atomic query occurs, and we embed for the non-atomic query with the least index on the path. If the query is not a simple yes-no question, we embed the appropriate standard decomposition e-scenario; otherwise we embed the standard e-scenario for the main connective of the affirmative answer to the query. The final result is an atomic e-scenario for the principal question assuming that the conditions specified by the Embedding Theorem are met.

---

[5] The procedure sketched above is, of course, not the only one possible. An interesting variant does not require fixing Y in advance. The difference lies in "building" Y step by step: once we arrive at a risky query, we choose a certain disjunction, D, of all the direct answers to the query that is not in $\mathsf{d}Q$, and we produce an atomic e-scenario for the query relative to D.

12.2 Transforming an e-scenario into an atomic one

For a procedure of systematic embedding operating with standard decomposition e-scenarios for connectives see Łupkowski Lupkowski (2010b). A computer program that generates atomic e-scenarios for simple yes-no questions was written in Prolog by Leszczyńska-Jasion.[6] A rule-based approach to systematic embedding (pertaining to the propositional case) is presented in Wiśniewski (2004a).

When dealing with the first-order level we operated with the language $\mathcal{L}^?_{fom}$. The expressive power of this language, however, is rather limited. Its vocabulary includes one-place predicates, identity, and individual constants, but no n-place predicates and function symbols. Moreover, the erotetic part comprises, besides whether-questions, only existential which-questions and general which-questions. But nothing essential would change if we extended the vocabulary with n-place predicates and/or function symbols, and the erotetic part with new categories of questions – assuming that the conditions ($\mathbf{sc}_1$), ($\mathbf{sc}_2$) and ($\mathbf{sc}_3$) specified in section 5.4.1 of Chapter 5 would still be fulfilled. As for semantics, we only relied upon some basic facts concerning entailment, mc-entailment and erotetic implication which are due to the meaning of propositional connectives in Classical Logic.

---

[6] The program can be downloaded from:
http://kognitywistyka.amu.edu.pl/intquestpro/

13

E-scenarios and Problem Solving

The concept of e-scenario was introduced in order to model some aspects of effective problem solving. Its applicability, however, has occurred to be wider. E-scenarios are useful tools in the area of cooperative answering, in a modelling of interrogator's hidden agenda, and in an analysis of the Turing Test.[1] Some aspects of question answering can be modelled by means of e-scenarios as well.[2]

In this Chapter we concentrate on problem solving.

13.1 Two kinds of problem decomposition

One of the crucial principles which govern effective problem solving is the following:[3]

(**DP**) (*Decomposition Principle*): *Decompose a principal problem (PP) into simpler sub-problems (SPs) in such a way that solutions to SPs can be assembled into an overall solution to PP.*

When we consider a problem definite enough to be adequately expressed by a question, its decomposition amounts, generally speaking, to finding an appropriate collection of auxiliary questions. A decomposition can be *static*, that is, resulting in finding a set of mutually independent auxiliary questions such that once *all* of them are answered, the initial problem is resolved.[4] Yet, a more interesting case is that of *dynamic* decomposition that comes in *stages*: the consecutive auxiliary questions (which constitute the sub-goals of the next stage) depend on how the previous requests for information have been fulfilled. The main goal, determined by the initial problem, remains unchanged, but sub-goals are processed in a goal-directed way. Moreover, the erotetic decomposition principle:

---

[1] Cf. Łupkowski (2010a), Urbański and Łupkowski (2010), Łupkowski (2010c), Łupkowski (2011).
[2] See Łupkowski (2012), Łupkowski (2013), Wiśniewski (201xa).
[3] I owe this formulation of **DP** to Mariusz Urbański (see Bolotov et al. (2006)).
[4] As for IEL, static decomposition is modelled in terms of reducibility of questions to sets of questions (see section 7.6.2 of Chapter 7).

(**EDP**) (*Erotetic Decomposition Principle*): Transform a principal question into auxiliary questions in such a way that: (a) consecutive auxiliary questions are dependent upon previous questions and, possibly, answers to previous auxiliary questions, and (b) once auxiliary questions are resolved, the principal question is resolved as well.

is obeyed until the initial problem becomes solved.

13.2 Dynamic decomposition via e-scenarios

13.2.1 Preliminary e-scenarios

Our claim is: when faced with a problem-solving task, it is *advisable* to build a *preliminary* e-scenario for the question that expresses the problem just considered. The background of the e-scenario is provided by items of information which are regarded as relevant to the case. The preliminary e-scenario can then be dynamically transformed in response to information gradually collected and by using the mechanisms of contraction and/or embedding.

An example can help to clarify the above claim.

Suppose that we aim at resolving the problem expressed by:

Where did Andrew leave for: Paris, London, or Rome?

and it is known, int. al., that:

$$\text{Andrew left for Paris, London or Rome.} \tag{13.1}$$

$$\text{If Andrew flew by Air France, then he left for Paris.} \tag{13.2}$$

$$\text{If Andrew did not fly by Air France, then he did not leave for Rome.} \tag{13.3}$$

$$\text{Andrew left for London if and only if he flew by BA or Rynair.} \tag{13.4}$$

A possible preliminary e-scenario is displayed in Figure 13.1.[5] For conciseness, we represent the set of initial premises (i.e. the set comprising the sentences (13.1) – (13.4)) by [**IP**].

A preliminary e-scenario provides information about possible ways of solving the principal problem: it shows what additional data should be collected and when they should be collected. The instructions provided are *conditional*: if one receives answer A to query Q, query Q^* should be asked next, if, however, one receives answer B to Q, query Q^{**} is the next one, etc. What is important, the e-scenario provides the appropriate instruction for *any of* the direct answers to the query: for each direct answer there is an instruction what to do next.

Let us stress: the principal question together with items of knowledge regarded as relevant to the case do not *uniquely determine* what is the preliminary

---

[5] As a matter of fact, the e-scenario depicted in Figure 13.1 is the compressed counterpart of the already analysed (see section 9.2.2 of Chapter 9) e-scenario displayed in Figure 9.2.

13.2 Dynamic decomposition via e-scenarios

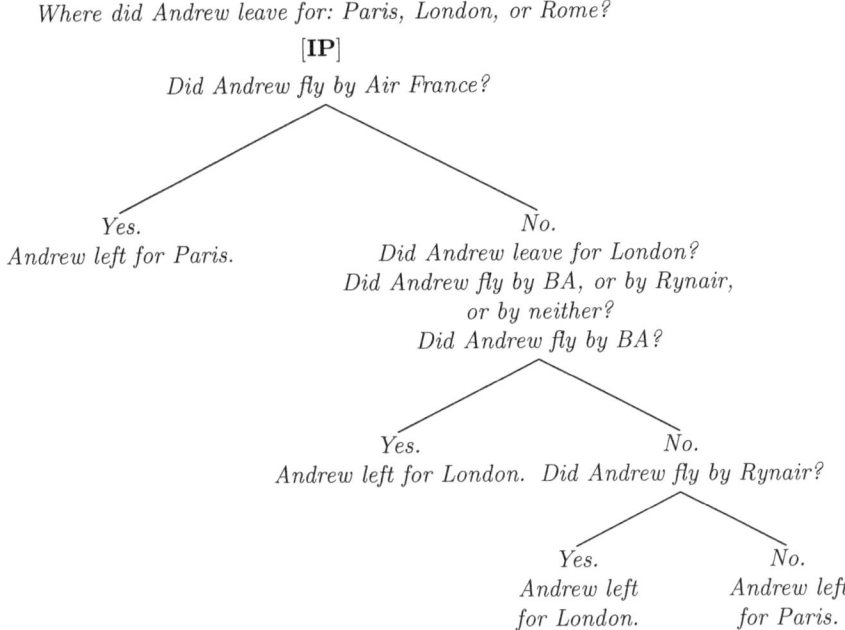

Fig. 13.1. *A preliminary e-scenario.*

e-scenario. In practice, it is wise to start with a preliminary e-scenario which has a relatively small number of queries. Moreover, one has to take into consideration that questions differ as to the "costs" incurred in order to obtain answers, where the costs are co-determined by such factors as the amounts of effort and/or time needed for obtaining an answer, data costs and/or charges, etc. It is advisable to use as queries only questions which are less "costly" than the principal one. And, last but not least, there must be good reasons to believe that answers to queries are available by accessible means.

A preliminary e-scenario is, in a sense, superfluous. The execution of the scenario is supposed to proceed from top to bottom: one attempts to resolve the first query and then, depending on the answer received, moves to the query recommended by the e-scenario as the next one, and so forth. However, instructions based on answers different from those which have been actually received (or hypothetically assumed; see below) will not be activated.

13.2.2 From query resolution to contraction

Looking from the formal point of view, success in resolving a query amounts to the contraction of the e-scenario executed by the just received answer to the query.

As an illustration, suppose that the first query of the e-scenario displayed in Figure 13.1 has been answered with "Andrew flew by Air France". The result of contraction by the above answer is an e-derivation ending with "Andrew left for Paris"; the endpoint is a direct answer to the principal question. Moreover, this answer is entailed by d-wffs which precede it in the derivation and thus is true *provided that* all the preceding d-wffs of the derivation are true.

Now suppose that the answer "Andrew did not fly by Air France" has been received. By contraction, the preliminary e-scenario transforms into the e-scenario displayed in Figure 13.2. Although originating in the previous one, this is a *new* e-scenario, which is supposed to be executed accordingly.

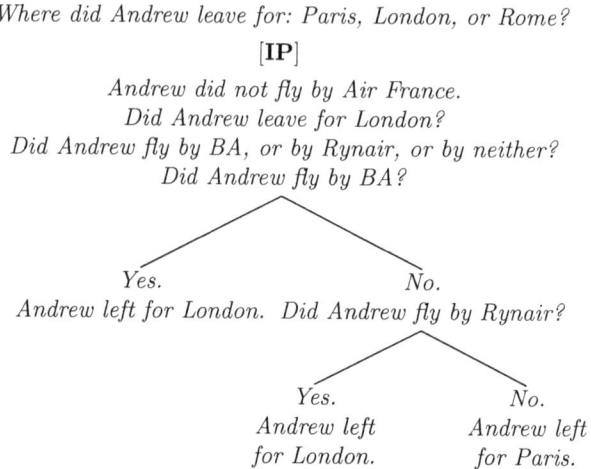

Fig. 13.2. *The e-scenario resulting by contraction.*

However, asking and successfully answering the first query of the scenario presented by Figure 13.2 can also be viewed as an execution of the instruction already present in the preliminary e-scenario displayed in Figure 13.1.

13.2.3 Embedding as a rescue option

A query conceivably resolvable at a reasonable cost can occur unanswerable by available means. When this happens, an advisable way out is to use the mechanism of embedding.

For instance, suppose that an inquirer executing the e-scenario displayed in Figure 13.2 faces a problem with the query "Did Andrew fly by BA?". An option is to embed a prospective e-scenario for the troublemaking query into the e-scenario which involves the query.

Assume that the e-scenario presented by Figure 13.3 is prospective enough.

The result of embedding it into the e-scenario depicted in Figure 13.2 is presented in Figure 13.4.

The e-scenario depicted in Figure 13.4 is supposed to be executed, starting with its first query, i.e. "Does Andrew prefer convenience over savings?". If the answer received is affirmative, contraction by this answer gives an e-derivation whose conclusion is: "Andrew left for London". If, however, the answer received

13.2 Dynamic decomposition via e-scenarios 167

Did Andrew fly by BA?
Andrew flew by BA if he prefers convenience over savings;
otherwise he did not fly by BA.
Does Andrew prefer convenience over savings?

```
            Yes.                       No.
     Andrew flew by BA.        Andrew did not fly by BA.
```

Fig. 13.3. *An e-scenario to be embedded into the e-scenario displayed in Figure 13.2.*

Where did Andrew leave for: Paris, London, or Rome?
[IP]
Andrew did not fly by Air France.
Andrew flew by BA if he prefers convenience over savings;
otherwise he did not fly by BA.
Did Andrew leave for London?
Did Andrew fly by BA, or by Rynair, or by neither?
Did Andrew fly by BA?
Does Andrew prefer convenience over savings?

```
            Yes.                       No.
     Andrew flew by BA.        Andrew did not fly by BA.
     Andrew left for London.   Did Andrew fly by Rynair?

                                  Yes.           No.
                              Andrew left     Andrew left
                              for London.     for Paris.
```

Fig. 13.4. *The result of embedding of the e-scenario displayed in Figure 13.3 into the e-scenario depicted in Figure 13.2.*

is negative, by contraction one arrives at the e-scenario displayed in Figure 13.5.[6]

A successful execution of the e-scenario depicted in Figure 13.5 gives, by contraction, either an e-derivation of "Andrew left for London" or an e-derivation of "Andrew left for Paris". Both are direct answers to the principal question. The answer actually arrived at is true provided that all the d-wffs preceding it at the path are true.

---

[6] The scenario involves the auxiliary question "Did Andrew fly by BA?" which has been already answered. However, this makes no harm since the question is not a query. In order to avoid such effects we would have to complicate the definition of contraction; sometimes already answered questions imply further questions, so an already answered non-query can be safely deleted only if it does not perform this role.

Where did Andrew leave for: Paris, London, or Rome?
[IP]
Andrew did not fly by Air France.
Andrew flew by BA if he prefers convenience over savings;
otherwise he did not fly by BA.
Andrew does not prefer convenience over savings.
Andrew did not fly by BA.
Did Andrew leave for London?
Did Andrew fly by BA, or by Rynair, or by neither?
Did Andrew fly by BA?
Did Andrew fly by Rynair?

```
         ╱      ╲
      Yes.       No.
   Andrew left  Andrew left
   for London.  for Paris.
```

Fig. 13.5. *The result of contraction of the e-scenario displayed in Figure 13.4 by the negative answer to 'Does Andrew prefer convenience over savings?'.*

13.2.4 Other rescue options and gains from a failure

Embedding can help, but there is no guarantee of success. What if some of the "new" queries remain unresolved? There are two possible rescue options. The first is to *backtrack* the already performed embedding and then embed another e-scenario for the troublemaking query. The second amounts to performing further embedding(s) without backtracking.

Let us note that persistent failures in resolving a query need not be tantamount to a complete failure. One can contract by an *only hypothetically accepted* answer to a troublemaking query and then try to proceed further. If one successfully proceeds with the consecutive queries recommended, the outcome carries information of the following kind: the endpoint provides a right solution to the initial problem *on condition that* the hypothetically accepted answers to the troublemaking queries or query are right. The added value of such outcome lies in an identification of knowledge gaps. However, if we rely on an answer (to a query) which is not supported by a sufficient evidence, it is likely that at some further point an unsound and/or unresolvable query will be the recommended one.

Remarks. At each stage of the process sketched above, with the exception of the last one, an e-scenario is executed. The consecutive e-scenario is dependent upon the result of execution of the previous one. Note that it is the preliminary e-scenario that is being transformed. As a consequence, the following desirable property is retained: each path of an intermediate scenario leads to an answer to the principal question. Thus the process as a whole is goal-directed, and the sub-goals are processed in a goal-directed way. Recall that any e-scenario has the golden path property[7] and describes a search plan with no "dead ends": the plan copes with any direct answer to a query.

However, a warning is in order. In the last stage of the process a solution to the initial problem emerges as the endpoint of the e-derivation arrived at. The solution is either a direct answer to the only query of the derivation or is entailed by some preceding d-wffs of the derivation. But it cannot be said that once the process is successfully completed, a "true" or "right" solution is already found. Mere entailment is not enough: there must be good reasons to believe that the premises involved and/or the answer to the query are true.[8]

13.2.5 Fine-tuning and systematic embedding

A preliminary e-scenario can be transformed in reaction to a success/failure in resolving a query. However, both embedding and contraction are formal operations which can be performed on a preliminary e-scenario prior to its execution. It is a rational strategy to estimate in advance both the chances of answering queries and costs of answering them. If the former are low and/or the latter are high, there is a possibility of "fine-tuning" the preliminary e-scenario by embedding. The idea is to begin questioning after having in one's disposal an e-scenario whose first query is answerable by available means at reasonable costs. The consecutive queries to be dealt with can be adjusted, if needed, in further steps. In some cases even a kind of general adjustment can be recommended. For instance, if there are good reasons to believe that atomic queries are both more promising and less costly than other queries, it is reasonable to transform the preliminary e-scenario into an atomic one and then execute the atomic e-scenario just obtained. As we have shown in section 12.2 of Chapter 12, in the case of e-scenarios whose queries are quantifier-free there are procedures whose application produce an atomic e-scenario as the outcome. Furthermore, the standard decomposition e-scenarios pertaining to queries can always be embedded when needed. One can also systematically reduce the complexity of quantifier-free yes-no questions occurring as queries by embedding the standard e-scenarios for connectives. The decomposition e-scenarios and scenarios for connectives can also be used as means of transforming intermediate e-scenarios.

Neither fine-tuning nor mere transformations of e-scenarios are sufficient to solve a problem: we need answers to queries. However, there is one exception. It can be shown that when one asks a yes-no question based on a CPL-valid formula, builds the appropriate (determined by the main connective) standard

---

[7] Cf. section 9.4 of Chapter 9.
[8] When Classical Logic is the underlying logic of declaratives, any scenario for Q relative to X which involves a query being a simple yes-no question can be transformed into an e-scenario which has *any* direct answer to Q as an endpoint of some path. It suffices to repeat the yes-no question. As a result, a pair of contradictory sentences emerges at a path, so we can conclude with any direct answer to Q. However, one cannot have good reasons to believe – dialetheism aside – that contradictory sentences are both true.

e-scenario and then transforms it consecutively by embedding (subjected to some diagrammatic rules) without making any attempts to answer the emerging queries, the final outcome is an e-scenario whose paths leading to the negative answer are contradictory. This, by the Golden Path Theorem, amounts to the affirmative solution of the principal problem (for details see Wiśniewski (2004a)). In other words, a systematic reflection on possible ways of reaching alternative solutions is sufficient to establish the right solution: no information-gaining moves are needed. Similarly, no information-gaining moves are needed in order to build a Socratic proof: mere transformations of questions are sufficient. So IEL throws a new light on the old idea of analyticity of logic.

Index

$A, B, C, D, \ldots$, 18
$A(x/c)$, 21
Ax, 21
$Q, Q^*, Q_1, \ldots$, 18
$[dQ]$, 43
Δ_B, 143
$\Sigma^{\Delta,\circledast}_{[\mathbf{s},\mathbf{s}_k]}$, 144
$\Sigma^{\Delta_B,\circledast}_{[\mathbf{s},\mathbf{s}_k]}$, 144
$\Sigma^{\ominus}_{[\mathbf{s},\mathbf{s}_{k+1}]}$, 148
$\Sigma_{[\mathbf{s},\mathbf{s}_k,B]}$, 143
$\Sigma_{[\mathbf{s},\mathbf{s}_m]}$, 141
$\underaccent{\smile}{\zeta}_{[\mathbf{t}]}$, 147
$\circledast$, 143
$\ominus$, 147
$\delta_{[\mathbf{g}]}$, 142
ℓ, 26
$\epsilon_{[\mathbf{s}]}$, 142
$\gamma_{[\mathbf{s}]}$, 142
$\hat{\zeta}_{[\mathbf{t}]}$, 147
$\leftrightarrow$, 18
$\Diamond$, 20
$\mathbb{E}^*$, 94
$\mathbb{G}^*$, 98
$\mathbf{E}$, 60
$\mathbf{E}_{\mathcal{L}^?_{fom}}$, 60
$\mathbf{E}_{\mathcal{L}^?_{cpl}}$, 60
$\mathbf{Im}$, 67
$\mathbf{S}(Ax)$, 22
$\mathbf{U}(Ax)$, 22
$\mathcal{D}_\mathcal{L}$, 25
$\mathcal{L}_{fom}$, 21
$\mathcal{L}^?_{fom}$, 21
$\mathcal{L}_\mu$, 20
$\mathcal{L}^?_\mu$, 20
$\mathcal{L}^?_{S4}$, 30

$\mathcal{L}^?_{\vdash cpl}$, 22
$\mathcal{L}_{cpl}$, 18
$\mathcal{L}_{cpl}$ sequent, 22
$\mathcal{L}^?_{cpl}$, 18
$?(\Phi)$, 23
$?A$, 20
$?[A_{|n}]$, 81
$?\mathbf{S}(Ax)$, 21
$?\mathbf{U}(Ax)$, 21
$?\pm|A,B|$, 19
$?\{A_1, \ldots, A_n\}$, 17
$\mathtt{PPres}Q$, 41
$\mathtt{Pres}Q$, 41
$\mathtt{d}Q$, 18
$\mathtt{faq}_\Delta$, 143
$\mathtt{ids}_\Delta$, 143
$\models_\mathcal{L}$, 33
$\models$, 37
$\models_{\mathcal{L}^?_{cpl}}$, 30
$\hat{\models}_\mathcal{L}$, 26
$\neg$, 18
$\rho_{[\mathbf{g}]}$, 142
$\rightarrow$, 18
$\Box$, 20
$\subset$, 26
$\vdash$, 22
$\vdash \ell$, 26
$\vee$, 18
$\wedge$, 18
$\widehat{\Sigma}_{[\mathbf{s},\mathbf{s}_m]}$, 142
$\zeta_{[\mathbf{s}]}$, 142
$'$, 28
$deg(A)$, 152
ng, 22
$;$, 90
$\&$, 22

172   Index

CTR, 148
EMB, 145
S4, 30
S4-model, 31

$\models$, 37

answers
  affirmative, 19
  corrective, 45
  direct
    to questions of $\mathcal{L}^?_{fom}$, 21, 22
    to questions of $\mathcal{L}^?_\mu$, 20
    to questions of $\mathcal{L}^?_{\vdash cpl}$, 23
    to questions of $\mathcal{L}^?_{cpl}$, 19
  direct to e-formulas, 18
  direct to NLQs, 18
  eliminative, 44
  just-complete, 43
  negative, 19
  partial, 43
  possible, 14
  principal possible, 14
approach to questions
  "define within", 13
  "enrich with", 14
  independent meaning thesis, 14
  paraphrase, 13
  semi-reductionistic, 16
atoms, 152
auxiliary question
  of an e-derivation, 111
  of an e-scenario, 113

background of an e-scenario, 113
basic sequent, 96

compactness
  of entailment, 85
  of mc-entailment, 70
contraction, 148
Contraction Theorem, 148
CPL, 18
CPL-valid sequent, 91
CPL-valuation, 30

d-wffs, 14
  compound, 152
  degree of, 152
  of $\mathcal{L}^?_{fom}$, 21
  of $\mathcal{L}^?_\mu$, 20
  of $\mathcal{L}^?_{\vdash cpl}$, 23

  of $\mathcal{L}^?_{cpl}$, 18
  quantifier-free, 152
  valid, 85
decomposition principle, 163

e-derivation (erotetic derivation), 110
e-formula, 14
e-scenario
  atomic, 153
  complete, 127
  compressed counterpart of, 123
  concise, 124
  imperative counterpart of, 124
  in the canonical form, 122
  incomplete, 127
  information-picking, 134
  non-atomic, 153
  pure, 128
  standard
    for a universal quantifier, 130
    for an existential quantifier, 130
    for conjunction, 128
    for disjunction, 129
    for equivalence, 130
    for implication, 128
    for negation, 128
e-scenario (erotetic search scenario)
  as a labelled tree, 116
  as a family of e-derivations, 113
elimination, 34
embedding, 145
  systematic, 160
Embedding Theorem, 145
entailment (multiple-conclusion)
  in a language, 33
entailment (single-conclusion)
  in a language, 26
erotetic calculus, 94
erotetic decomposition principle, 103
erotetic formula, 14
erotetic implication, 67
  analytic, 77
  auxiliary d-wffs, 67
  falsificationist, 72
  implied question, 67
  implying question, 67
  pure, 77
  regular, 76
  strong, 76
erotetic inference, 49
  of the first kind, 49, 50
    conditions of validity, 51

valid, 65
  of the second kind, 50
    conditions of validity, 52
    valid, 73
erotetic rules, 94
evocation of questions, 60

generation of questions, 63
Golden Path Theorem, 116

imperative counterpart of an e-scenario, 124
Inferential Erotetic Logic, 49
  the logical basis of, 56
initial d-wffs
  of an e-derivation, 111
initial premises
  of an e-derivation, 111
  of an e-scenario, 113

just-sufficiency, 18

language $\mathcal{L}_{fom}$, 21
language $\mathcal{L}^?_{fom}$
  semantics, 31
  syntax, 21
language $\mathcal{L}_\mu$, 20
language $\mathcal{L}^?_\mu$
  semantics, 30
  syntax, 20
language $\mathcal{L}^?_{S4}$
  semantics, 30
language $\mathcal{L}^?_{\vdash cpl}$
  semantics, 27
  syntax, 22
language $\mathcal{L}^?_{cpl}$
  semantics, 29
  syntax, 18

mc-entailment, 33
MiES, 25
Minimal Erotetic Semantics, 25
model of $\mathcal{L}^?_{fom}$, 32
  normal, 32

narrowing down, 35
NLQ, 14

partition
  admissible, 26, 27
  of $\mathcal{L}^?_{fom}$, 32
  of $\mathcal{L}^?_{S4}$, 31

of $\mathcal{L}^?_{\vdash cpl}$, 29
of $\mathcal{L}^?_{cpl}$, 30
improper, 27
of a language, 26
of the set of d-wffs, 25
proper, 27
path of an e-scenario, 113
ppa, 14
presuppositions, 39
principal question of an e-scenario, 113
prospective presupposition, 40

query
  of an e-derivation, 112
  of an e-scenario, 114, 116
    atomic, 153
    risky, 158
question
  conditional yes-no, 19
  atomic yes-no, 153
  binary, 43
  choice, 9
  conjunctive, 19
  contingent, 39
  delimited-condition, 8
  non-factual, 84
  normal, 41
  open-condition, 5
  proper, 42
  quantifier-free, 153
  regular, 41
  risky, 38
  safe, 38
  self-rhetorical, 42
  simple yes-no, 19
  topically-oriented, 10
  whether, 152
questions
  as erotetic formulas, 18
  of $\mathcal{L}^?_{fom}$
    existential which, 21
    general which, 21
    whether, 21
  of $\mathcal{L}^?_\mu$, 20
  of $\mathcal{L}^?_{\vdash cpl}$, 23
  of $\mathcal{L}^?_{cpl}$, 18

reducibility, 87
riskiness, 38

safety, 38
sentence, 21

sentential function, 21
Socratic proof in $\mathbb{E}^*$, 96
Socratic transformation in $\mathbb{E}^*$, 95
soundness
    of a question, 38
    relative, 42

standard decomposition e-scenarios, 131, 133
synthetic tableau, 136

wff, 18
    α, 28
    β, 28
wffs of $\mathcal{L}$, 110

References

Asher, N. and Lascarides, A. (2003), *Logics of Conversation*, Cambrige University Press, Cambridge.

Batens, D. (2007), Content guidance in formal problem solving processes, *in* O. Pombo and A. Gerner, eds, 'Abduction and the Process of Scientific Discovery', Centro de Filosofia das Ciencias da U. de Lisboa, Lisbon, pp. 121–156.

Belnap, N. D. (1969), Åqvist's corrections-accumulating question sequences, *in* J. Davis, P. Hockney and W. Wilson, eds, 'Philosophical Logic', Reidel, Dordrecht, pp. 122–134.

Belnap, N. D. and Steel, T. P. (1976), *The Logic of Questions and Answers*, Yale University Press, New Haven.

Bolotov, A., Łupkowski, P. and Urbański, M. (2006), Search and check: Problem solving by problem reduction, *in* A. Cader and et. al, eds, 'Artificial Intelligence and Soft Computing', Academic Publishing House EXIT, Warszawa, pp. 505–510.

Bromberger, S. (1992), *On What we Know We Don't Know. Explanation, Theory, Linguistics, and How Questions Shape Them*, The University of Chicago Press and CSLI, Chicago/Stanford.

Buszkowski, W. (1989), 'Presuppositional completeness', *Studia Logica* **48**, 23–44.

De Clercq, K. (2005), *Logica in communicatie*, Academia-Bruylant, Louvain-la-Neuve.

De Clercq, K. and Verhoeven, L. (2004), 'Sieving out relevant and efficient questions', *Logique et Analyse* **185-188**, 189–216.

Enqvist, S. (2010), 'Interrogative belief revision in modal logic', *Journal of Philosophical Logic* **38**, 527–548.

Ginzburg, J. (1995), 'Resolving questions I', *Linguistics and Philosophy* **16**, 459–527.

Ginzburg, J. (2011), Questions: logic and interactions, *in* J. van Benthem and A. ter Meulen, eds, 'Handbook of Logic and Language. Second Edition', Elsevier, Amsterdam/Boston/Heidelberg, pp. 1133–1146.

Ginzburg, J. (2012), *The Interactive Stance: Meaning for Conversation*, Oxford University Press, Oxford.

References

Ginzburg, J. and Sag, I. (2000), *Interrogative Investigations. The Form, Meaning and Use of English Interrogatives*, Vol. 123 of *CSLI Lecture Notes*, CSLI, Stanford.

Grobler, A. (2006), *Metodologia nauk*, Znak & Aureus, Kraków.

Grobler, A. (2012), 'Fifth part of the definition of knowledge', *Philosophica* **86**, 33–50.

Grobler, A. and Wiśniewski, A. (2005), Explanation and theory-evaluation, *in* R. Festa, A. Aliseda and J. Peijnenburg, eds, 'Cognitive Structures in Scientific Inquiry. Essays in Debate with Theo Kuipers', Rodopi, Amsterdam/New York, pp. 299–310.

Groenendijk, J. (2011), Erotetic languages and the inquisitive hierarchy, *in* J. van der Does and C. Dutilh Novaes, eds, 'This is not a Festschift – Festschrift for Martin Stokhof', https://sites.google.com/site/inquisitivesemantics/papers-1/publications.

Groenendijk, J. and Roelofsen, F. (2009), Inquisitive semantics and pragmatics, *in* J. Larrazabal and L. Zubeldia, eds, 'Meaning, Content and Argument, Proceedings of the ILCLI International Workshop on Semantics, Pragmatics and Rhetoric', University of the Basque Country Publication Service, pp. 41–72.

Groenendijk, J. and Stokhof, M. (1997), Questions, *in* J. van Benthem and A. ter Meulen, eds, 'Handbook of Logic and Language', Elsevier & The MIT Press, Amsterdam/Cambridge, pp. 1055–1125.

Groenendijk, J. and Stokhof, M. (2011), Questions, *in* J. van Benthem and A. ter Meulen, eds, 'Handbook of Logic and Language. Second Edition', Elsevier, Amsterdam/Boston/Heidelberg, pp. 1059–1132.

Hamblin, C. L. (1958), 'Questions', *The Australasian Journal of Philosophy* **36**, 159–168.

Harrah, D. (1981), The semantics of question sets, *in* D. Krallman and D. Stickel, eds, 'Zur Theorie der Frage', Gunter Narr Verlag, Tubingen, pp. 36–45.

Harrah, D. (1997), On the history of erotetic logic, *in* A. Wiśniewski and J. Zygmunt, eds, 'Erotetic Logic, Deontic Logic, and Other Logical Matters. Essays in Memory of Tadeusz Kubiński', Wydawnictwo Uniwersytetu Wrocławskiego, Wrocław, pp. 19–27.

Harrah, D. (2002), The logic of questions, *in* D. Gabbay and F. Guenthner, eds, 'Handbook of Philosophical Logic, Second Edition', Vol. 8, Kluwer, Dordrecht/Boston/London, pp. 1–60.

Higginbotham, J. and May, R. (1981), 'Questions, quantifiers, and crossing', *The Linguistic Review* **1**, 41–80.

Hintikka, J. (1976), *Semantics of Questions and the Questions of Semantics*, Vol. 28 of *Acta Philosophica Fennica*, North-Holland, Amsterdam.

Hintikka, J. (1978), Answers to questions, *in* H. Hiż, ed., 'Questions', Reidel, Dordrecht, pp. 279–300.

Hintikka, J. (1999), *Inquiry as Inquiry: A Logic of Scientific Discovery*, Kluwer, Dordrecht/Boston/London.

Hintikka, J. (2007), *Socratic Epistemology: Explorations of Knowledge-Seeking by Questioning*, Cambridge University Press, Cambridge.

Hintikka, J., Halonen, I. and Mutanen, A. (2002), Interrogative logic as a general theory of reasoning, in D. Gabbay, R. Johnson, H. Ohlbach and J. Woods, eds, 'Handbook of the Logic of Argument and Inference', North-Holland, Amsterdam, pp. 295–337.

Krifka, M. (2011), Questions, in K. von Helsinger, C. Maieborn and P. Portner, eds, 'Semantics. An International Handbook of Natural Language Meaning. Vol. II', Mouton de Gruyter, Berlin/New York, pp. 1742–1785.

Kubiński, T. (1960), An essay in the logic of questions, in 'Proceedings of the XIIth International Congress of Philosophy (Venetia 1958)', Vol. 5, La Nuova Italia Editrice, Firenze, pp. 315–322.

Kubiński, T. (1971), *Wstęp do logicznej teorii pytań*, Państwowe Wydawnictwo Naukowe, Warszawa.

Kubiński, T. (1980), *An Outline of the Logical Theory of Questions*, Akademie-Verlag, Berlin.

Kuipers, T. and Wiśniewski, A. (1994), 'An erotetic approach to explanation by specification', *Erkenntnis* **40**, 265–284.

Leśniewski, P. (1997), *Zagadnienie sprowadzalności w antyredukcjonistycznych teoriach pytań*, Wydawnictwo Naukowe Instytutu Filozofii UAM, Poznań.

Leśniewski, P. (2000), On the generalized reducibility of questions, in J. Nida-Rumelin, ed., 'Rationality, Realism, Revision. Proceedings of the 3rd International Congress of the Society for Analytical Philosophy', Walter de Gruyter, Berlin/New York, pp. 119–126.

Leśniewski, P. and Wiśniewski, A. (2001), 'Reducibility of questions to sets of questions: some feasibility results', *Logique et Analyse* **173–175**, 93–111.

Leszczyńska, D. (2004), 'Socratic proofs for some normal modal propositional logics', *Logique et Analyse* **185–188**, 259–285.

Leszczyńska, D. (2007), *The Method of Socratic Proofs for Normal Modal Propositional Logics*, Wydawnictwo Naukowe UAM, Poznań.

Leszczyńska-Jasion, D. (2008), 'The method of socratic proofs for modal propositional logics: K5, S4.2, S4.3, S4M, S4F, S4R and G', *Studia Logica* **89**, 371–405.

Leszczyńska-Jasion, D. (2009), 'A loop-free decision procedure for modal propositional logics K4, S4 and S5', *Journal of Philosophical Logic* **38**, 151–177.

Leszczyńska-Jasion, D. (2013), Erotetic search scenarios as families of sequences and erotetic search scenarios as trees: two different, yet equal accounts, Technical report, Department of Logic and Cognitive Science, Institute of Psychology, Adam Mickiewicz University in Poznań.

Leszczyńska-Jasion, D., Urbański, M. and Wiśniewski, A. (2013), 'Socratic trees', *Studia Logica*, DOI: 10.1007/s11225-012-9404-0 .

Łupkowski, P. (2010*a*), Cooperative answering and inferential erotetic logic, in P. Łupkowski and M. Purver, eds, 'Aspects of Semantics and Pragmatics of Dialogue. SemDial 2010, 14th Workshop on the Semantics and Pragmatics of Dialogue', Polish Society for Cognitive Science, Poznań, pp. 75–82.

Łupkowski, P. (2010*b*), Erotetic search scenarios and problem decomposition, in D. Rutkowska, J. Kacprzyk and A. Cader, eds, 'Some New Ideas and Research Results in Computer Science', Academic Publishing House EXIT, Warszawa, pp. 202–214.

Łupkowski, P. (2010c), *Test Turinga: perspektywa sędziego*, Wydawnictwo Naukowe UAM, Poznań.

Łupkowski, P. (2011), 'A formal approach to exploring the interrogator's perspective in the Turing Test', *Logic and Logical Philosophy* **20**, 139–158.

Łupkowski, P. (2012), Erotetic inferences and natural language dialogues, *in* 'L & C 2012. Proceedings of the Logic and Cognition Conference, Poznań, 17-19 May, 2012', Poznań, pp. 39–48.

Łupkowski, P. (2013), Cooperative posing of questions, *in* V. Punčochař and P. Švarny, eds, 'The Logica Yearbook 2012', College Publications, London, pp. 79–90.

Meheus, J. (1999), 'Erotetic arguments from inconsistent premises', *Logique et Analyse* **165-167**, 49–80.

Meheus, J. (2001), 'Adaptive logic for question evocation', *Logique et Analyse* **173-175**, 135–164.

Minică, Ş. (2011), Dynamic Logic of Questions, PhD thesis, Institute for Logic, Language and Computation, University of Amsterdam.

Negri, S. and von Plato, J. (2001), *Structural Proof Theory*, Cambridge University Press, Cambridge.

Olsson, E. J. and Westlund, D. (2005), 'On the role of the research agenda in epistemic change', *Erkenntnis* **65**, 165–183.

Peliš, M. (2011), Logic of Questions, PhD thesis, Charles University in Prague, Faculty of Arts, Department of Logic.

Peliš, M. and Majer, O. (2011), Logic of questions and public announcements, *in* N. Bezhanishvili, S. Löbner, K. Schwabe and L. Spada, eds, 'Logic, Language, and Computation 8th International Tbilisi Symposium on Logic, Language, and Computation', Vol. 6618 of *Lecture Notes in Computer Science*, Springer, Berlin/Heidelberg, pp. 145–157.

Rasiowa, H. and Sikorski, R. (1963), *The Mathematics of Metamathematics*, Państwowe Wydawnictwo Naukowe, Warszawa.

Shoesmith, D. J. and Smiley, T. J. (1978), *Multiple-conclusion Logic*, Cambridge University Press, Cambridge.

Smullyan, R. (1968), *First-order Logic*, Springer, Berlin/New York.

Urbański, M. (2001a), 'Remarks on synthetic tableaux for classical propositional calculus', *Bulletin of the Section of Logic* **30**, 194–204.

Urbański, M. (2001b), 'Synthetic tableaux and erotetic search scenarios: extension and extraction', *Logique et Analyse* **173-174-175**, 69–91.

Urbański, M. (2002a), 'Synthetic tableaux for Łukasiewicz's calculus L3', *Logique et Analyse* **177-178**, 155–173.

Urbański, M. (2002b), *Tabele syntetyczne a logika pytań*, Wydawnictwo Uniwersytetu Marii Curie-Skłodowskiej, Lublin.

Urbański, M. and Łupkowski, P. (2010), Erotetic search scenarios: Revealing interrogator's hidden agenda, *in* P. Łupkowski and M. Purver, eds, 'Aspects of Semantics and Pragmatics of Dialogue. SemDial 2010, 14th Workshop on the Semantics and Pragmatics of Dialogue', Polish Society for Cognitive Science, Poznań, pp. 67–74.

van Benthem, J. and Minică, Ş. (2012), 'Toward a dynamic logic of questions', *Journal of Philosophical Logic* **41**, 633–669.

Vanderveken, D. (1990), *Meaning and Speech Acts, 2 vols.*, Cambridge University Press, Camridge.
Wiśniewski, A. (1986), Generowanie pytań przez zbiory zdań, PhD thesis, Institute of Philosophy, Adam Mickiewicz University in Poznań.
Wiśniewski, A. (1989), The generating of questions: a study of some erotetic aspects of rationality, *in* L. Koj and A. Wiśniewski, eds, 'Inquiries into the Generating and Proper Use of Questions', Wydawnictwo Uniwersytetu Marii Curie-Skłodowskiej, Lublin, pp. 91–155.
Wiśniewski, A. (1990*a*), 'Implied questions', *Manuscrito* **13**, 23–38.
Wiśniewski, A. (1990*b*), *Stawianie pytań: logika i racjonalność*, Wydawnictwo Uniwersytetu Marii Curie-Skłodowskiej, Lublin.
Wiśniewski, A. (1991), 'Erotetic arguments: A preliminary analysis', *Studia Logica* **50**, 261–274.
Wiśniewski, A. (1994*a*), 'Erotetic implications', *Journal of Philosophical Logic* **23**, 174–195.
Wiśniewski, A. (1994*b*), 'On the reducibility of questions', *Erkenntnis* **40**, 265–284.
Wiśniewski, A. (1995), *The Posing of Questions: Logical Foundations of Erotetic Inferences*, Kluwer, Dordrecht/Boston/London.
Wiśniewski, A. (1996), 'The logic of questions as a theory of erotetic arguments', *Synthese* **109**, 1–25.
Wiśniewski, A. (1997a), Some foundational concepts of erotetic semantics, *in* M. Sintonen, ed., 'Knowledge and Inquiry. Essays on Jaakko Hintikka's Epistemology and Philosophy of Science', Rodopi, Amsterdam/New York, pp. 181–211.
Wiśniewski, A. (1999), 'Erotetic logic and explanation by abnormic hypotheses', *Synthese* **86**, 295–309.
Wiśniewski, A. (2001), 'Questions and inferences', *Logique et Analyse* **173–175**, 5–43.
Wiśniewski, A. (2003), 'Erotetic search scenarios', *Synthese* **134**, 389–427.
Wiśniewski, A. (2004*a*), 'Erotetic search scenarios, problem-solving, and deduction', *Logique et Analyse* **185–188**, 139–166.
Wiśniewski, A. (2004*b*), 'Socratic proofs', *Journal of Philosophical Logic* **33**, 299–326.
Wiśniewski, A. (201x*a*), Answering by means of questions in view of inferential erotetic logic, *in* J. Meheus, E. Weber and D. Wouters, eds, 'Logic, Reasoning and Rationality', Springer.
Wiśniewski, A. (201x*b*), Semantics of questions, *in* S. Lappin and C. Fox, eds, 'Handbook of Contemporary Semantics. Second Edition', Wiley-Blackwell.
Wiśniewski, A. and Shangin, V. (2006), 'Socratic proofs for quantifiers', *Journal of Philosophical Logic* **35**, 147–178.
Wiśniewski, A., Vanackere, G. and Leszczyńska, D. (2005), 'Socratic proofs and paraconsistency: A case study', *Studia Logica* **80**, 431–466.